P9-CJD-150

Blind Watchers of the Sky

THE PEOPLE AND IDEAS THAT SHAPED OUR VIEW OF THE UNIVERSE

Rocky Kolb

HELIX BOOKS

ADDISON-WESLEY PUBLISHING COMPANY

Reading, Massachusetts Menlo Park, California New York

Don Mills, Ontario Harlow, England Amsterdam Bonn

Sydney Singapore Tokyo Madrid San Juan

Paris Seoul Milan Mexico City Taipei

Many of the designations used by manufacturers and sellers to distinguish their products are claimed as trademarks. Where those designations appear in this book and Addison-Wesley was aware of a trademark claim, the designations have been printed in initial capital letters.

Library of Congress Cataloging-in-Publication Data

Kolb, Rocky.
Blind watchers of the sky : the people and ideas that shaped our view of the universe / Rocky Kolb.
p. cm.
Includes bibliographical references and index.
ISBN 0-201-48992-9
1. Cosmology. 2. Solar systems. I. Title
QB981.K689 1996 95–41438
523.1—dc20 CIP

Jacket design by Jean Seal
Text design by Karen Savary
Set in 11-point Minion by Pure Imaging

1 2 3 4 5 6 7 8 9-MA-0099989796
First printing, February 1996

To my parents

Contents

Foreword

Who has not looked up at the night sky and asked those eternal questions: Who are we? How did we get here? Where are we going? The first records that we find on the misty origins of science are these cosmological questions generated by the star-studded great vault of the sky. Modern cosmology, joined to particle physics, is at the most tentative edges of our scientific worldview, and this book is yet another effort to bring the mind-boggling summary to a general audience. The author once dreamt that he was a partial derivative! This so shook him up that he vowed to bring enlightenment without mathematics. He does it by using storytelling, history, and a Woody Allen-esque solemnity that will keep you giggling and learning.

Edward (Rocky) Kolb has been the director of the astrophysics group at Fermilab and is responsible for replenishing and reinvigorating the connection between the inner space of particle physics as carried out in the Fermilab atom-smashers and the outer space of early universe cosmology that deals with the universe when it was a hot, bubbling soup of fundamental particles. It takes a certain style to make and keep the connections lively. Rocky's self-effacing exuberance, his hesitant militancy, and his neoconservative iconoclasty have all helped to create an environment of nervous excitement. It really adds a lot to an experiment on neutrino masses to learn that a high neutrino mass would imply the future collapse of the universe (the Big Squeeze). But Kolb's book is astronomy from ancient Greece to modern brouhaha of black holes, the microwave background, *und so weiter.*

What I find most gratifying about this popularization is that it also teaches what science is—a tortured assembly of contrary qualities: of skepticism and rationality, of freedom and revolution, of passion and aesthetics, and of soaring imagination and trained common sense. The advance of astronomy, like that of so many other pieces of science, is full of errors, uncertainty, false directions...but also a gradual approach to consensus, punctuated by rare breakthroughs and *Eureka!* moments. Kolb tells the history through the very human qualities of the scientist as each one fashions and places his or her stone in the arch of our understanding.

Blind Watchers of the Sky will certainly make a great movie or twelve-part television series with billions and billions of viewers. In these times of a dizzying pace of change in our end-of-century society driven by science and technology, there is, as never before, a need to raise the level of public understanding of how science works. Kolb serves us well.

As you close the book with that fine mixture of regret, pleasure, and profit, one overriding question will remain with you: Should you let your offspring marry an astronomer?

Leon M. Lederman
Illinois Institute of Technology

Preface

In the sixteenth century it was common to publish a work anonymously, and only if the book was a success would the identity of the author be slipped to a few friends and colleagues. Although I write scientific papers and technical books under the name Edward W. Kolb, the use of "Rocky" for this book is not an attempt at anonymity, for most friends and colleagues already call me Rocky (for a reason I have never been able to understand). Rather, the use of the pseudonym was to serve as a constant reminder to myself that the audience for this book is not the usual target audience for whom I write. If this schizoid ploy succeeds, a reader with no technical knowledge of astronomy, cosmology, or physics should be able to enjoy the book.

The purpose of the preface has changed a bit since it was first introduced. Its original function was to provide an opportunity for an author to apologize for the inadequateness of the book, and to plead that it really wasn't meant for publication in the first place, but a well-meaning friend or a colleague pried away a manuscript intended only for private amusement and forced it on a publisher. (If such a scenario seems fanciful, look into the history of the publication of the famous book of Nicolas Copernicus!)

Later, the preface evolved into the place to thank those who offered help or encouragement. But that is not the purpose of this preface; indeed, the list of those deserving mention is of sufficient length to merit a separate section at the end of the book.

The sole purpose of this preface is to reveal what the book is about. This book is about cosmology, the study of the universe. In 1543, Copernicus imagined that someone without training in astronomy could read his book and start calculating planetary positions, for on the title page of the first edition of *The Revolutions* appears the promotional kicker:

> [In the book] You also have very convenient tables from which you can most easily calculate [planetary positions] for all time. Therefore buy, read, use!

It is no longer feasible, however, to present in a single book a complete cosmological worldview. Modern science is far too complicated for anyone to master a field from one book, particularly one that is non-technical. But even without complete mastery of the technical details, anyone can develop an appreciation of what we know about the universe, and how we came to know it.

It is my experience that the scientific method is easily confused with the scientific process. Although cold, hard, perhaps inhuman observational and experimental facts are the basis on which scientific theories are ultimately judged, the process of developing new ideas in science is a very human endeavor. As with any enterprise involving people, the process is very personal. There might be a neat characterization of the scientific method, but the scientific process is as varied as human personalities. The process of science involves heroic ideas as well as its share of stupidity.

This book is not meant to be a comprehensive history of the development of cosmology. Nor is it meant to be a comprehensive astronomy text, or even a complete exposition of the big bang. It is a personal view of some of the chapters of the story of the development of our picture of the universe as viewed from the perspective of a modern cosmologist. For although the tools of modern cosmology include some of the most complicated scientific instruments ever built, the basic spirit is the same as that which guided Kepler, Galileo, and Newton. After all, the greatest tool of discovery is the human imagination. The history of cosmology is more of a story of the triumph of the human spirit and imagination than it is the development of a scientific discipline.

The book is divided into three sections. The first section recounts the development of our view of the universe when it was thought to consist principally of the solar system. The story as recounted here starts at the beginning of modern astronomy in the sixteenth and seventeenth centuries. Since it begins with modern astronomy, it does not discuss in detail the work of the Greeks, or the oriental astronomers, or the mathematicians and astronomers in the Arabic world. A history of their contributions must be found elsewhere.

The second part of the book involves the extension of our worldview to include the stars and the Milky Way. The most important part of that story involves the long, arduous process of determining the distances to planets, stars, and galaxies.

The final section extends our worldview into the larger universe, and the development of our modern model of the universe, the big bang. This is a tale of twentieth-century physics and astronomy. Like all stories in science, it ends with questions.

References, a selected bibliography, and suggestions for further reading are provided at the end of the book. Also included at the end of the book is a glossary/appendix, "The Devil in the Details," elaborating on a few of the technical aspects for those interested in a deeper understanding of some of the astronomy used in the chapters. Key words for topics discussed in the appendix appear in **boldface** type in the text.

Above all, this book is intended for those curious about the universe. Although the origin and structure of the universe can be truly understood only by the few hundred cosmologists throughout the world who devote their lives to its study, the basic ideas can be understood by anyone with the desire to know. Cosmologists have a duty to explain the universe to the public. You have a right to know; after all, *it's your universe too!*

The most important thing I want to communicate is the joy of cosmology. Everyone who has studied the heavens feels at times like the Greek astronomer Ptolemy, who, in the second century of the Christian Era, wrote:

> I know that I am mortal by nature, and ephemeral; but when I trace at my pleasure the windings to and fro of heavenly bodies I no longer touch Earth with my feet, but

> stand in the presence of Zeus himself and take my fill of ambrosia, food of the gods.

There have been moments when it felt as if my feet left Earth. It is a wonderful experience. I hope this book conveys some sense of that feeling.

Therefore buy, read, enjoy!

Warrenville, Illinois
January 1996

O crassa ingenia. O cœcos cœli spectatores.
Oh thick wits. Oh blind watchers of the sky.

Tycho Brahe, 1573

ONE

Eyes on the Skies

There are many beautiful and wondrous things to see in the universe, and to discover them we simply have to gaze into the dark night sky. Astronomers tell us that. But what do they know? Many astronomers spend a great deal of their lives in secluded parts of the world on mountaintops breathing the thin, rarefied air found at high altitudes and looking through large telescopes. Like most people, I live near sea level in the environs of a large city; in my case, Chicago. If I go out on one of those rare evenings that are clear and not too cold and look at the sky for a few minutes, I really can't see all that much to greatly excite my imagination. The most prominent sights in my night sky are the planes approaching or leaving O'Hare Field, just twenty miles away. Some might say they can just look into the sky and see a thousand points of light. I can't.

Those who live at higher altitudes, in drier climates, and away from city lights have a much better view of the night sky. But can they see all that much more? Indeed they have a better view of the features of the Moon, they can see more stars, possibly they can even begin to perceive that stars have different colors, and they should be able to notice that there are regions of the sky (like the Milky Way) that appear to be fuzzy. But what can they really see that will help them comprehend the universe?

If the lucky people with a clear view of the sky watch it over the course of an entire night, they will notice that the stars seem to move, circling around the North Star. If they enjoyed the first evening's view, they might decide to come back for more. After a few weeks of patient observing they would notice that some objects in the sky seem to have peculiar motions. Not only does the Sun travel from east to west every day, but it rises and sets at different locations on the horizon. The Moon also appears to wander around the sky, and it seems to change shape, sometimes appearing round, sometimes looking like a crescent. A few watchers of the sky will become so addicted to sky gazing that it will become a habit, and soon the sky will grow familiar to them. After a few months these persistent watchers of the sky will be rewarded with the realization that there are some peculiar starlike objects that do not remain fixed in relation to the other stars, but seem to wander across the rest of the sky. They will then understand why these objects are called "planets," after the Greek word for "wanderer." If they are like most people, they will search for the comfort that comes from finding regularity in the motions of objects in the sky.

Every couple of decades watchers of the sky will notice a new star appear and then slowly fade. In a lifetime of study, our inveterate watchers of the sky might see one or two comets, a few lunar eclipses, and, if they are very lucky, a solar eclipse. Finally, every couple of centuries a generation of watchers will be rewarded with the spectacular sight of a star suddenly blazing into view and becoming one of the brightest lights in the sky for a year or so before again fading out of sight.

Of those who look into the sky, some will find answers, but most will find questions. Sky watching has led people to ask some of the most profound questions about our universe. Where did it come from? How big is it and how old is it? Where is the edge, the center? What is beyond? And in this century, invariably they will ask what occurred before there was a universe. Such queries about the universe are some of the most fundamental questions ever posed. In reply, they will be told, and they will believe, that the answers to their questions are written upon the sky, if only they are clever enough to be able to decipher them.

For some deep reason, every culture has asked questions about the universe and has employed watchers of the sky to provide answers. In the twentieth century, the study of such questions is part of a scientific discipline known as *cosmology.* Cosmologists study the origin and the structure of the universe. The word cosmology is derived from the Greek κόσμος (cosmos). In this context κόσμος does not mean enormous, or immense. Rather, it is the Greek word for "order." Cosmologists attempt to bring an order to our view of the universe.

Modern cosmologists use physical laws as tools to understand an apparently complex and mysterious universe. The strategy is straightforward: learn the laws of physics by performing laboratory experiments, and explain the observed universe on the basis of these laws. The reward for the effort is a *model* of the universe. Telescopes and other modern instruments generally do not lead to qualitatively different information; they mostly enable us to refine and quantify the types of measurements we can make with our eyes. We need a model to make the most of the meager information we can acquire. Every civilization has had models of the universe. The presently accepted picture of the universe is known as the *big-bang model.*

Before discussing the big bang, or any other model for that matter, it is important to understand exactly why it is necessary to have a model in hand when confronting the universe. The first reason is the sheer enormity of the number of objects in the universe. On a clear night, on a mountain away from city lights, about 2,500 stars, one moon, and as many as five planets are visible to the unaided eye. With a good pair of binoculars, and enough patience to examine the entire sky, more than 160,000 stars can be seen. If the mountaintop happens to have an observatory on it, a photographic plate at the focus of a moderate-sized telescope could image over seventeen million stars if enough time is taken to survey the entire sky. If the observatory happens to house a telescope with a five-meter mirror—say, Mount Palomar Observatory outside of San Diego, California—then with modern electronic detectors substituted for photographic plates it is possible to image many more than a billion stars in the visible part of the sky. Of course, the large telescopes would reveal more than stars in our own galaxy, the Milky Way. With large telescopes one begins to see external galaxies and other objects beyond our galaxy.

Even more objects can be found by observing the sky in wavelengths of the **electromagnetic spectrum** that cannot be seen by the human eye. Modern astronomical observations are performed over the entire range of the spectrum, from long-wavelength radio waves to short-wavelength gamma rays. In fact, many objects can be seen only when studying the sky with wavelengths of light that are not "visible."

But even limiting ourselves to objects emitting energy in wavelengths visible to the eye, we can find billions of things up there.

A truly democratic approach to astronomy would be to study every object equally. The trouble is that even working ten hours a day, every day, for forty years (and even astronomers take a day off every decade or so), one would need to study two objects per second to examine a billion things. Of course, no astronomer works this way, because not every object in the sky tells us something new about the universe. To make sense out of what we see, we must sort through a multitude of objects to find those that are useful to study further. To do this, we need a model of the universe.

To illustrate, let's pick a region of the sky and look at it in progressively finer detail. We start by looking at a familiar part of the sky in one of the eighty-eight constellations—say, the constellation Virgo. The Sun is in the Virgo constellation in the autumn, so six months later in the spring, Virgo is overhead at night. If we look toward the center of the

A LOOK AT THE UNIVERSE

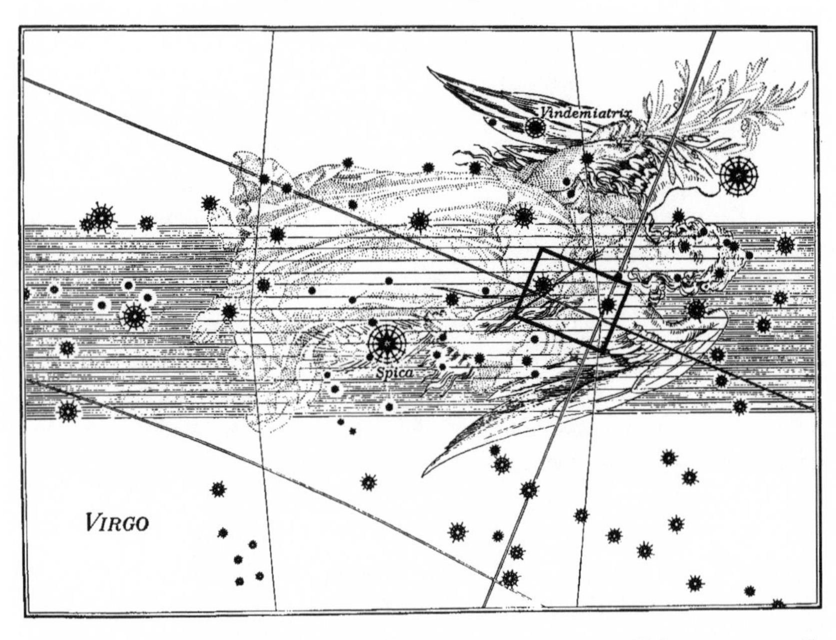

The Virgo constellation as depicted by Bayer in 1603. We will focus on a small region of the constellation shown by the box. It contains two easily visible stars named Porrima and Zaniah.

constellation we can see with the unaided eye two stars, named Porrima and Zaniah.

Now let's turn a large telescope to that region of the sky. The telescope greatly extends our vision and allows us to see objects not visible to the unaided eye. But the light from dim objects is retained by the eye for only a brief instant. If we could collect the light for a longer period of time, we could further extend our vision. The use of photography gives us this ability, because photographic film will collect the imprint of the light for as long as it is exposed. Long-exposure photography will freeze on film what our eyes cannot see. If we remove our eye from the eyepiece of the telescope and replace it with a photographic plate, then we can take a picture of the sky in the Virgo constellation. The most striking thing in the picture is how staggeringly many objects there are. Not

A CLOSER LOOK AT THE UNIVERSE

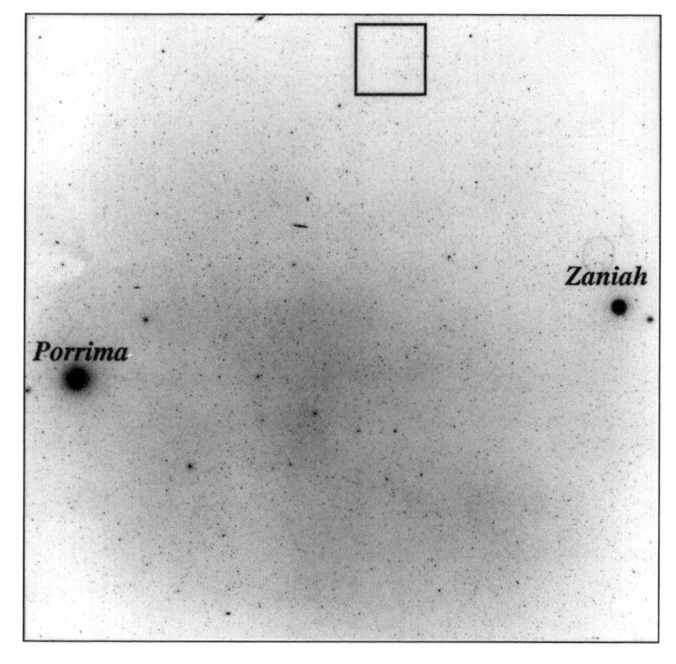

A region of the sky in the Virgo constellation observed by Caltech astronomers using the Mount Palomar Observatory. The star on the left is Porrima, and the star on the right is Zaniah. The picture was taken in 1955 with a one-hour exposure at the focus of a seventy-inch telescope. We will take an even closer look at the area indicated by the inner box.

only do we see many more stars with the aid of the telescope, but we see galaxies and other objects outside of our galaxy. But how do we know that the galaxies are very distant? How do we know what we are seeing? To answer such questions we need a model of the universe, for without a model we are swamped by the sheer number of things we see. With a model of the universe we can discover that of the thousands of objects on the picture, some are particularly noteworthy.

All objects in the universe are interesting and beautiful in their own right, but not all objects have the potential to lead to a greater appreciation of nature. Without a model, all of the thousands of points of light on the photograph look alike. But with a model, we can grasp the significance of the fact that some of the marks on the picture look like extended smudges: they are galaxies external to our own. With a model, we also discover that one of the objects in the box at the top of

AN EVEN CLOSER LOOK AT THE UNIVERSE

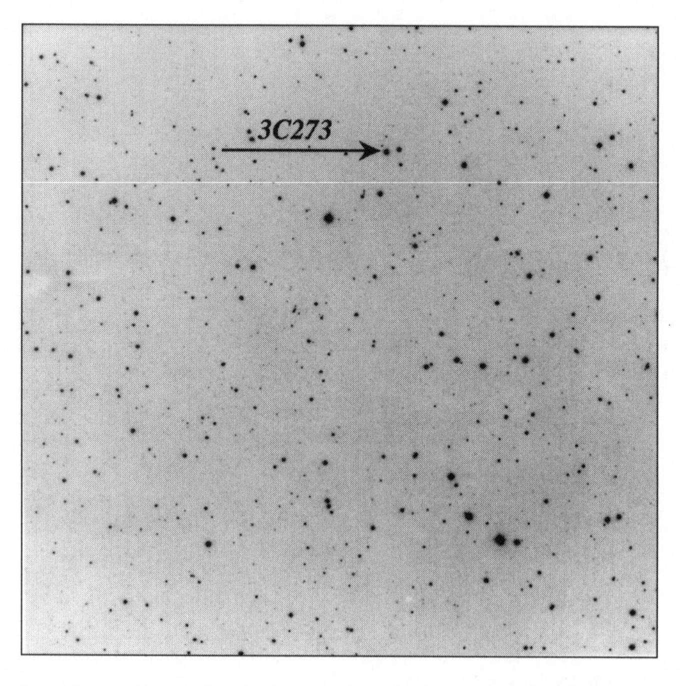

A blowup of the region of the sky denoted by the box in the previous photograph. The region indicated by the arrow is the first quasar discovered, 3C273, the 273rd entry in the third Cambridge catalog of celestial sources of radio emission.

the picture is peculiar. When we blow up the region of the box and study it in still finer detail, we notice that one of the dots (not the largest and not the smallest) is also a source of radiation in the long-wavelength, radio region of the electromagnetic spectrum. When we study that particular object even more closely, we find that although it looks like a dim star, it is at an enormous distance from us. In fact, that one dot, known by the unromantic name of 3C273, was the first quasar discovered. It was identified as a quasar in 1962 by Maarten Schmidt, but its image had appeared on photographs like this for decades, and it was also noted in radio surveys of the sky. In fact, it is not all that dim, and can be seen with the aid of a small telescope, appearing as a faint blue star.

The story of the discovery of something as fundamental as quasars demonstrates that the ability to see something does not naturally lead to an understanding of its significance. Only with the aid of a cosmological model is it possible to sort through the bewildering mass of information and identify the points of light that might be special. The story illustrates that the secrets of nature are not easily elucidated. Sadly for the astronomer, nature does not provide for arrows in the sky to point out the interesting objects; they must be ferreted out by hard work. It also raises the question of which other small dots on the picture are the keys to unlock the mysteries of the universe. Astronomers are haunted by the thought that discoveries are before their eyes but they are too blind to see them.

The second reason why a model of the universe is crucial is that there is not much we can learn about the things in the universe. This statement may seem puzzling, given that we have just seen that so many objects exist in the universe. However, there is simply not much information we can obtain directly about any one object. For instance, observing a bright object with the unaided eye, we can determine its color, its apparent brightness, its apparent **angular size**, and its relative position with respect to other objects in the sky. Of course, we can also determine any change in these observables. This is not much information with which to put together a model of the universe.

To my mind, the most remarkable aspect of astronomy is found not in the sky but on Earth, where a species has developed the curiosity to look into the sky and wonder. Perhaps the origin of sky watching was driven by practical considerations, like predicting the seasons to know when to sow and when to reap, or when a nearby river would rise. Other possibilities have been postulated for the genesis of astronomy. But it really doesn't matter whether astronomy began for the Darwinian reason that it led to the increased probability of food production and hence a greater chance for survival, or for the Marxist explanation that knowledge of astronomy gave shamans special power that could be used to subjugate the masses, or the capitalist rationale that astronomy developed for the crass commercial purposes of improving navigation. For whatever purpose it began, astronomy today has clearly progressed far beyond any reason other than to fulfill a deep longing in us to understand our universe. The longing is strongest in those who devote their lives to astronomy, cosmology, or physics, but some vestige of the longing must reside deep within us all as part of human nature.

The task of understanding the universe is greater than might be imagined, for divining the workings of the universe by watching the sky is an arduous job. Indeed, developing a model of the universe is one of the most difficult intellectual challenges ever undertaken. But the rewards are correspondingly great. That on an insignificant planet in orbit about an average star in a typical galaxy we look into the sky, not in fear but with the bold confidence that we can comprehend something so much larger than we are, tells us something important about ourselves. The very fact that so many people have devoted their lives to this task, and we have developed as great an understanding as we have, is one of those inexplicable things like art or music that should give us all pride in the accomplishments of our species.

There are many stories to be told about cosmology. One story is our present understanding of the origin and structure of the universe. Equally interesting is the story of the steps in this long process and the people involved in obtaining the knowledge. Perhaps even more remarkable than the story of what we know, or the people and steps involved in the development of that knowledge, is the fact that the story is told at all—that we have a desire, indeed a need, to know.

As a cosmologist, I feel that all the stories should be told, for they have a real significance. We cherish and honor many aspects of life in our culture—philosophy, religion, art, music, literature, and so on. These can all be thought of as possible expressions of our search for an answer to the question, "What is our place in the universe?" Cosmology is concerned with the answer to an even more basic question "What is the universe?" Today, and throughout history, our attempts to answer the first question have been influenced by our perception of the answer to the second question.

The human search for an understanding of the universe has taken us far. When our ancestors first looked at the sky, it was with blind eyes.

Since then our vision has grown sharper, but we know we have yet to see it all. Although today we look harder and farther than we have ever looked, with all the instruments of modern science, and with all of the imagination and courage we can muster, we are still blind watchers of the sky.

PART 1

The Solar System

TWO

Smashing the Celestial Spheres

Twenty thousand light-years away, just over twelve million years ago, a new star was born from a mix of dust and gas and debris of long-dead stars. Although the birth of a star is a remarkable phenomenon, it is not uncommon. A new star is born in our galaxy every few years. In a vast galactic recycling program, stars continually form from material in interstellar space, and the more massive stars return some of the matter to the interstellar medium through stellar winds that blow gas from the surface of the star, or through more violent, explosive processes. Stellar births and deaths are part of the life cycle of a galaxy. The most important aspect of this particular star is that it was more massive than a typical star, about fifteen times as massive as our Sun. Because of its large mass, it was destined to lead a short, hot life, and meet a violent end.

This star lived the first part of its life producing energy by the fusion of hydrogen into helium, just as our Sun is doing now. As the energy flowed out from the center, it pushed on the outer layers of the star and exactly balanced the force of gravity trying to pull the outer layers into the center. The large mass of the star resulted in a very strong gravitational pull, and in order to provide a correspondingly larger push, the star had to produce more power than our Sun, so it burned hotter and faster. It consumed all of the hydrogen in its core in just twelve million years. (Our smaller Sun, with less mass to support, does not have as great an appetite for fuel. The Sun has been burning hydrogen for nearly five *billion* years, and will continue to do so for another five billion years.) After the star had burned the hydrogen, it consumed the helium, the ashes of hydrogen burning. As the star aged, it burned hotter and hotter, and consumed fuel at an ever increasing rate. It took only a little over one hundred thousand years to burn all the helium in its core, and then it consumed the ashes of helium burning, carbon, and converted it to neon in just six thousand years. The consumption of the ashes of a previous fuel continued through the burning of neon to oxygen, oxygen to silicon, and silicon to iron—all this in only seven years.

The star now had a serious weight-control problem. Once it had burned the fuel to iron, it could not satisfy its need for energy production. Because iron is indigestible, it will not burn in nuclear reactions and produce energy. After the star had produced a core of just under two solar masses of iron, it could no longer produce the energy necessary to prevent the thirteen solar masses outside the core from crashing down on it. The core of the star collapsed under its weight until its center reached the density found in the interior of nuclei and could collapse no further. As the outer layers of the star continued to rain down on the dense core, a shock wave was produced deep within the star. The shock wave propagated out through the enormous star, blowing off the outer layers of the star in a giant explosion known as a **supernova**. For a brief time this exploding star was the brightest star in the galaxy—as bright, in fact, as all the rest of the stars in the galaxy combined!

When the star exploded, sometime around 18,000 B.C., humans in the Paleolithic era were just picking up sticks and scratching crude representations of their environment on cave walls. Perhaps around this time they first turned their sights to the sky to notice the stars. Of course, no one would have seen this star the instant it exploded. Since the star is twenty thousand light-years away, the light from the explosion would take twenty thousand years to reach Earth. Until the arrival of the light from the explosion, anyone looking in the direction of the constellation Cassiopeia would not have noticed this particular star at all. Even though the energy output of the star in the millennia before the explosion was ten thousand times greater than our Sun, it was just a little too far away to have been visible to the unaided eye. It would have been easily visible through binoculars (not readily available in the Paleolithic era), but it would have appeared as just one of thousands of red supergiant stars in the sky.

Finally, some of the light from the exploding star reached Earth on November 11, 1572. On that night a young Danish astronomer named Tycho Brahe[1] had just finished supper and was returning to his observatory for a typical night's work when, as astronomers often do, he happened to look at the sky. Although he was only twenty-six years old, since the age of thirteen he had been gazing at the heavens with an earnestness surpassed by very few before him (and very few after him). The heavens were as familiar to Tycho as the furniture in his bedroom. He stopped in his tracks, amazed to see a bright star in a region of the sky where before there had been none. A new star had appeared in the constellation Cassiopeia—a new star that outshone every other star in the night sky, a new star even brighter than Venus. Tycho could not believe his eyes. He was shocked to see the splendor of a new star in the heavens he knew so well. His astonishment was as great as yours might be if you were to wake up in the middle of the night and discover that a grand piano had appeared in the familiar confines of your bedroom. Perhaps thinking he had a bit too much wine with dinner, he quickly called for his servants to confirm what his eyes could scarcely believe.

All of Europe was astonished by the appearance of a new star. No one knew what it meant. The German painter George Busch warned that the new star augured the coming of all sorts of calamities, such as "inclement weather, pestilence, and Frenchmen."

Twenty-three years after Tycho's supernova, Shakespeare was to write, in *King Richard II*, "for violent fires soon burn out themselves." So it was for the supernova. When it first appeared, it blazed as the brightest star in the night sky, even visible during the day for weeks after its appearance. It slowly faded, however, and a year and a half later it had disappeared from sight. Although gone, it was not forgotten, for the appearance of the star left its imprint on Tycho, indeed on the future development of astronomy.

Exploding stars, or supernovae, are not rare. In galaxies like ours they occur on average once or twice per century. On a human time scale

[1] Tycho (pronounced *Tee-ko*) is the latinized form of Brahe's Danish first name, Tyge (pronounced *Tee-geh*). His family name, Brahe, is pronounced *Bra-hee.*

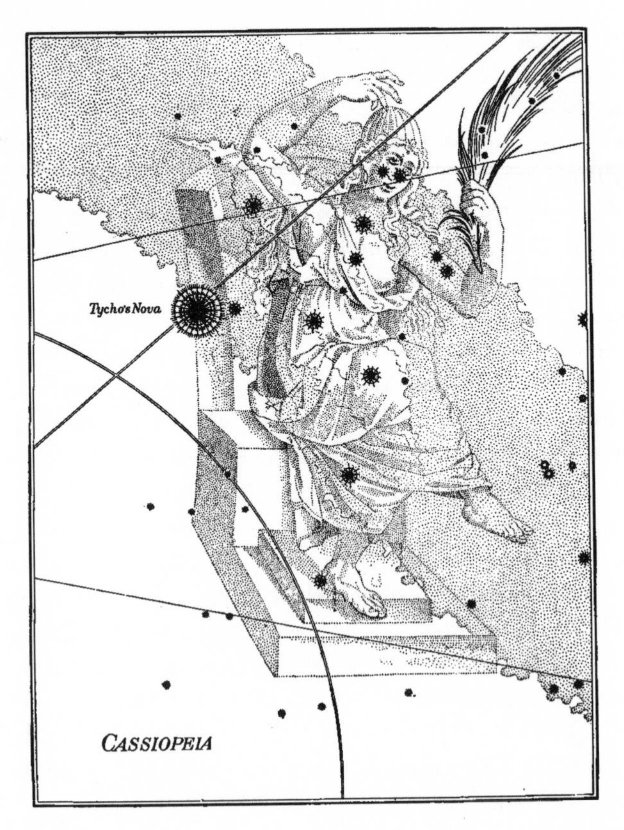

The position of Tycho's supernova in the constellation Cassiopeia.

a century is a long time, but for a ten-billion-year-old galaxy, a century is like the mere wink of an eye in a human life span. But this supernova was special. The light from this supernova not only briefly outshone the entire galaxy but fell on the eyes of Tycho Brahe, one of the greatest astronomers of all time. It reached Earth at a time when Western civilization had advanced enough for someone to grasp the significance of the new star.

An understanding of the birth and evolution of stars would not be possible until the twentieth-century synthesis of the sciences of thermodynamics, hydrodynamics, astronomy, atomic physics, nuclear physics, and elementary particle physics into modern astrophysics. Our understanding of the mechanism of supernova explosions is still not complete. Tycho and other astronomers of the sixteenth century had no idea what the new star was, but they were still astonished that something in the heavens had changed.

Whether or not we realize it, our cosmology, our view of the universe, has a powerful grip on us. The cosmological zeitgeist of sixteenth-century Europe was shaped by the teachings of the great Greek philosopher Aristotle, which was molded to fit into Christian dogma. The cosmology of the day was that the heavens, since they were created by God, were perfect and unchanging. Earth, although also created by God, suffered change and decay because it was corrupted by the presence of humans. It was inconceivable to an Aristotelian that a new star could appear in the sky in the immutable eighth sphere of the heavens.[2] Comets[3] were known to appear and disappear, but they were believed to be atmospheric phenomena, and within the corrupting influence of humankind. Historians knew that Pliny had noted that the Greek astronomer Hipparchus in the second century B.C. had registered the appearance of a new star, but the Aristotelians held that he had merely seen a comet.

How difficult it was for people to free themselves from the grip of this doctrine, or any other doctrine for that matter, is hard to appreciate. Tycho's supernova was not the only one to appear in the Christian Era. On July 4, in the year A.D. 1054, a supernova even brighter than the one of 1572 appeared in the constellation Taurus. This spectacular event was noted and studied by astronomers in China, Japan, Korea, and the Middle East, and even by the Navajo Indians in the American Southwest. European history is curiously silent about the appearance of this new star, however. Did the intellectual confines of an unchanging, perfect universe blind people to the significance of the event? Were they blind to the light, or blinded by the light? Perhaps our eyes cannot see what our mind is not prepared to accept. In 1006 a still brighter supernova had appeared. At its maximum brightness this supernova was as bright as a half-moon. Chinese astronomers reported that the new star shone so brightly that one could clearly see objects at night by its light. All European records

[2] The first seven spheres in the **Ptolemaic system** were thought to carry the planets going outward from the central Earth (the Moon, Venus, Mercury, the Sun, Mars, Jupiter, Saturn). The eighth sphere contained the stars.

[3] The word *comet* comes from the Greek κομη'τηζ, which means "the hairy one."

indicate that the supernova was believed to be a comet. Various monastic chronicles at the time record the following:

> ...in 1006 there was a very great famine and a comet appeared for a long time....
>
> ...and so at the same time, a comet, the sign of which always announces human shame, appeared in the southern regions, which was followed by a great pestilence throughout all the territories of Italy and Venice...
>
> ...three years after the king was raised to the throne, a comet with a horrible appearance was seen in the southern part of the sky, emitting flames this way and that....

The supernova was believed to be a comet, and the only thing to do in 1006 was to cower in the corner and await the coming of the inevitable plague, pestilence, war, or Frenchmen.

But the world of 1572 was a different place from the world of 1054 or 1006. Perhaps the life of the average person had not changed much in the intervening five centuries, but interspersed among the farmers, merchants, and soldiers were a new breed of people, people of exceptional talent and vision who would revolutionize our view of the universe. Nicholas Copernicus's revolutionary book that proposed removing Earth from the center of the universe had appeared only twenty-nine years previously. In 1572, when the supernova appeared, Johannes Kepler, who would eventually work with Tycho and succeed him as the greatest astronomer in Europe, was an eleven-month-old child in Germany, and an eight-year-old Galileo Galilei, who was destined to invent modern science, was playing in the streets of Pisa. The Renaissance had swept Europe, creating an intellectual atmosphere in which such people could flourish. In turn, Tycho, Kepler, and Galileo, and later Newton, would bequeath to the world a new way of looking at the universe.

Of course, not all astronomers thought the new star signaled a new astronomy. Many dismissed the apparition of 1572 as a tailless comet, some sort of atmospheric phenomenon, a condensation of the rising vapors of human sin, or even a sign of the wrath of God. But after a year of meticulous observations, Tycho Brahe understood that far removed from the Earth a new star had appeared. Few of Tycho's contemporaries believed him, because two millennia previously Aristotle had said that the heavens could never change. Impatient with those who had minds but would not think, and frustrated by those who possessed eyes but could not see, Tycho lashed out at them in the preface of his book on the new star:

> *O crassa ingenia. O cœcos cœli spectatores.*
>
> Oh thick wits. Oh blind watchers of the sky.

THE ASTRONOMER KING

One can immediately recognize an opera as having been composed by Giuseppe Verdi, although each opera has what Verdi described as its own *tinta e colorito,* tint and coloring. So it is with the lives of Copernicus, Tycho, Kepler, Galileo, Newton, and others who have shaped our view of the universe. In their work they were consumed with the same passion: to understand and comprehend and explore the universe. They were kindred spirits, but their lives, background, and personality were distinctly different. Scientific genius, like musical genius or artistic genius, comes in all shapes, sizes, and types. Each one had his individual tint and coloring.

Tycho Brahe was born in 1546, into one of the most prestigious families of Denmark. All four of Tycho's great-grandfathers, both his grandfathers, and his father had been members of the Danish Rigsraad, an elite twenty-member council who chose kings, ratified treaties, declared war, and defended the interests of the ruling class. In the land of Hamlet, Tycho's paternal grandmother was descended from the powerful families of Rosenkrantz and Gyldenstierne. Tycho's father was governor of the strategic Helsingborg Castle. At the age of two, Tycho was adopted (*kidnapped* is a more accurate word) by his childless uncle Jørgen Brahe, the vice-admiral of Denmark. Jørgen was a favorite of Fredrick II, king of Denmark. Fredrick must have felt a debt to young Tycho. Tycho's paternal grandfather, the first Tycho Brahe, was killed in the war that placed Fredrick I on the throne. His foster father not only was victorious in naval battles against the Swedes but personally saved the king's life when Fredrick II accidentally fell into the icy waters of Denmark near the royal castle in Copenhagen.[4] Jørgen's reward for the heroic rescue of his king was a fatal case of pneumonia. Tycho's family connections gave him access to the

[4] Several contemporary sources seem to imply that the king was inebriated (and perhaps Jørgen as well).

Tycho Brahe at forty, surrounded by the crests of the families of his illustrious ancestors.

inner corridors of power and influence. In his lifetime Tycho was more at home in a royal court than he ever could be in an academic setting, and he would entertain as many kings as professors.

Tycho's youth was typical of nobility. While still a twenty-year-old student, he fought a duel with his third cousin.[5] According to legend, the duel was not precipitated by an argument over a woman: rather, it

[5] Dueling seems to have been a major cause of death in the Brahe family. Jørgen's brother was killed in a duel, Tycho's second cousin was killed by his uncle in a duel, one of Tycho's third cousins killed someone in a duel, and one of Tycho's second cousins killed his first cousin (Tycho's second cousin) in another duel. There must have been a decidedly nervous edge to conversations at family dinners.

was a dinnertime argument over who was the better mathematician. Tycho both saved face and lost face in the duel: his honor was upheld, but his nose was sliced off. "Nose jobs" were somewhat primitive in the sixteenth century; Tycho had attached to his face a nose of gold fashioned by Denmark's finest craftsmen.

In the spirit of the times, his family planned a career for Tycho as a statesman. It was to be a career in which he could live the life of pleasure and luxury they felt he was entitled to because of his noble station. As part of his education, he was enrolled at the University of Copenhagen as a student in philosophy. But in 1559 Tycho was "starstruck" by a solar eclipse. It was only a partial solar eclipse, and not all that spectacular. What seemed to have struck the thirteen-year-old Tycho was not the appearance of the eclipse but the fact that an occurrence in the heavens could be *predicted* by mere mortals.

Tycho's life was never the same after he experienced the eclipse. The heavens are seductive. Many of us feel the powerful pull and the wonder instilled by a deep gaze into the dark night sky. Many noble or wealthy people have been fascinated by astronomy and contributed either money or time to it. Some have even become important astronomers. Tycho was in a position to pursue astronomy as an aristocratic dilettante. But the stars are harsh. Once they truly have hold of you they don't let go, and Tycho was at their mercy. Although he remained a nobleman, from the age of thirteen he was first and foremost an astronomer. In 1560 Tycho purchased the standard astronomy textbook of the time, *On the Spheres*, by Sacrobosco, and in 1561, at the age of fifteen, he obtained *Trigonometry*, the advanced book on mathematical astronomy by the great astronomer Regiomontanus.

The unseemly life of an astronomer was not part of the career path envisioned by his family. The Brahe family expected Tycho to be a royal administrator, governor, or admiral, eventually filling his ancestral seat on the Rigsraad. One can imagine the anguished cries of his family: "No child of mine is going to grow up to be an astronomer!" Tycho was sent to a private tutor, Soerensen Vedel, for the purpose of curing Tycho of this eccentric, disreputable vice of star gazing. For the first year Tycho was forced to read astronomy books at night by candle under his blankets and to hide quadrants and other astronomical instruments in his room. But Vedel, a greatly respected scholar in his own right, was a much better tutor than the family had bargained for. He soon realized that it was hopeless to try to drive the stars from the mind of Tycho. To his great credit he gave up, and in the end encouraged Tycho's interest in

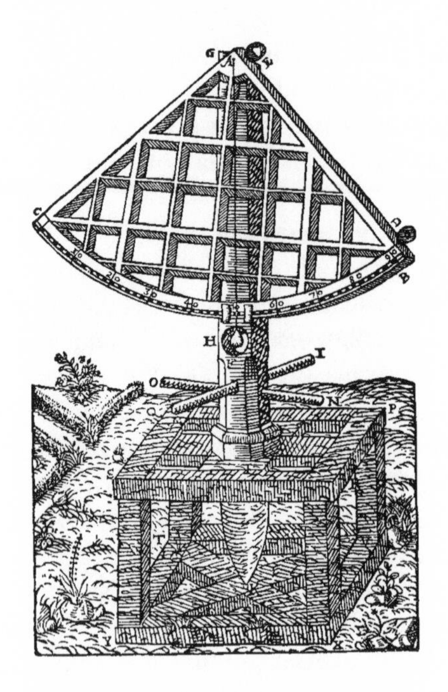

A drawing of a great quadrant built by Tycho in 1570. Constructed in Augsburg of oak and brass, the apparatus was over thirty feet in height and required the labor of forty men to move it. In his book Astronomiæ Instauratæ Mechanica *of 1598, describing this and other instruments, Tycho lamented that it was left derelict five years after construction: "Since men as a rule are more interested in worldly matters than in things celestial, they usually regard with indifference such happenings, which will perhaps be more harmful to them than they themselves realize."*

astronomy. In return, Vedel became Tycho's lifelong friend, adviser, and confidant.

Tycho continued his education at some of the best universities in Germany. While still a student he had the private resources and the ability to build astronomical instruments better than any others in the world. At the young age of twenty-six when the supernova of 1572 appeared, he was the greatest astronomer in Europe. The light from this supernova finally fell on the eyes of someone prepared to appreciate its significance.

✦ ✦ ✦

If the new star was in the vicinity of Earth, it should have an apparent motion with respect to the "fixed" stars. Although the stars have an apparent daily, or diurnal, motion, their relative positions do not change. The stars of the Big Dipper seem to rotate about Polaris, the North Star, but they always appear to have the same familiar shape. Astronomers throughout Europe set about to see whether the new star of 1572 moved with respect to the fixed stars. Some astronomers detected a spurious motion; most thought it remained fixed. But Tycho had at his disposal superior instruments. It would be another thirty-six years before the invention of the telescope, so astronomers had to employ a variety of means to determine the relative position of the stars. While other astronomers fashioned bits of string to measure the distance from the new star to other stars, Tycho could use a bronze quadrant over six feet in size. While others could only lash out at the problem with pieces of string, Tycho could bring to the battle the equivalent of a tank.

This was perhaps the first example of how an expensive, enormous piece of equipment can lead to new discoveries. Luckily Tycho had private resources available to construct the instruments. He could never have justified to a funding agency the cost of construction of his instruments. One can imagine congressional committees demanding to know exactly what discoveries would be made with it, or how he could justify using great quantities of bronze at a time when it was needed for cannons to defend the realm. Tycho would have been forced to say that *maybe* with new instruments he *might* be able to make more precise measurements, which *could* be important to settle some important scientific issue, which *perhaps* would arise in the future. Not all correct arguments sound convincing. Not all scientific projects have an immediate, spectacular payoff. But some do.

Armed with his instruments, Tycho was able to make very accurate determinations of the position of the new star, and he was able to show that the new star didn't move. His book about the new star, known by the appropriate title *The New Star*,[6] established Tycho at age twenty-six at the pinnacle of the astronomical world. Although Tycho was not the first, and certainly not the only, astronomer to study the new star of 1572, because of the quality of his observations it is justly known as Tycho's supernova.

Tycho was quick to parlay his fame into a position no astronomer had ever dreamt of, or would ever hope for again. King Fredrick II of Denmark was an enlightened monarch of the Renaissance, a true patron of the arts. He was impressed by Tycho's work and reputation and offered him various castles for his use, and the resources for the construction of an observatory and astronomical instruments. Displaying a degree of arrogance found only in genius or nobility (he happened to be both), Tycho ignored the offers from his king and left for a grand tour of

[6] The complete title is *Mathematical Contemplation of Tycho Brahe of Denmark on the New and Never Previously Seen Star Now First Observed in the Month of November in the Year of Our Lord 1572.* For some reason the title was shortened by historians.

Europe to visit emperors and astronomers, letting it be known that he "had quadrant—will travel" if anyone was interested in providing a suitable position for a prince who just happened to be the world's greatest astronomer. Of course, Tycho wanted to stay in Denmark, but he thought he could cut a better deal with the king if he generated (or appeared to generate) a few offers elsewhere. Since the beginning of recorded history, such brinksmanship has been an often practiced, but not very respected, part of the academic game. [7] It is one thing to play games with department chairs or deans, but it takes quite a different person to wheel and deal with a king.

Tycho's gamble paid off, for Fredrick offered Tycho something that seems incredible. Tycho was to be deeded the entire island of Hveen (now part of Sweden, but then owned by Denmark), all of the crown's tenants and servants on the island, as well as the rent and duty generated by everyone living on the island. And he was to construct at the state's expense a castle for his personal use, a chemical laboratory, a

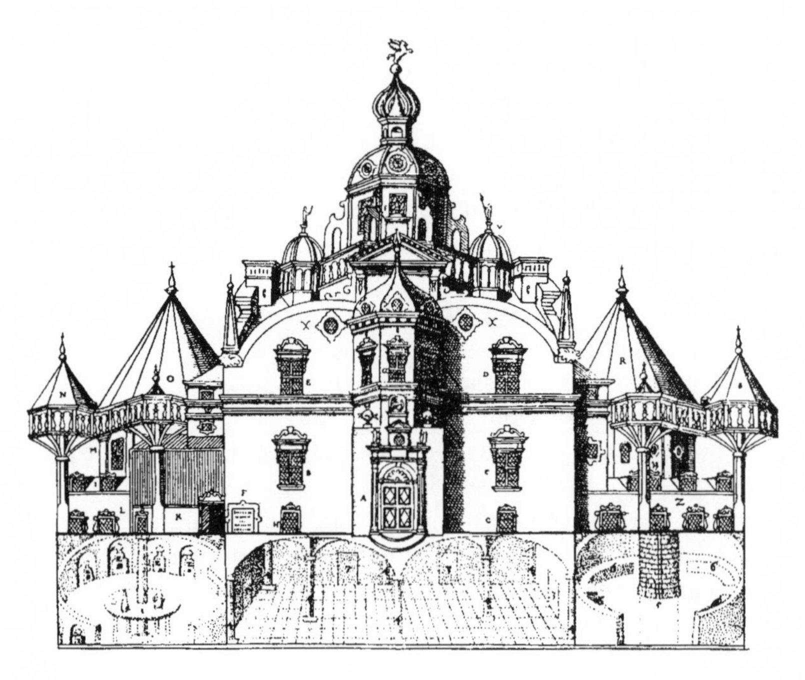

Named for Urania, the muse of astronomy, Uraniborg was finished in 1581.

[7] Of course, this has never been done by any of *my* colleagues.

printing press, a paper mill, a prison, a game preserve for his amusement, a library, an observatory the likes of which the world had never known, and the largest, most elaborate astronomical instruments ever constructed.

After receiving this staggering offer, Tycho kept the king on hold for only two weeks before accepting it. Tycho's observatory, Uraniborg, was constructed at the expense of 30 percent of the budget of Denmark, the modern equivalent of about $5 billion. Many of Fredrick's advisers thought him mad to underwrite Tycho's observatory.[8] Modern observatories, even modern particle accelerators, are bargains by comparison. In his elaborate castle he entertained princes and kings, as well as the occasional astronomer. Tycho was not only king of astronomers, but an astronomer king. Finally, an astronomer was receiving proper treatment!

THE NEXT DECIMAL PLACE

Often the development of science is perceived to run as a precision gear, with progress clicking along in a smooth, straight, continuous sequence of advances made by a series of scientific heroes. But science does not typically proceed in this clockwork manner. More often than not, the way science goes from point *A* to point *B* is by a random lurch through points *X*, *Y*, and *Z*. Even when great leaps of progress do occur, they only rarely come "out of the blue." Advances are nearly always preceded by years, decades, or even centuries of patient accumulation of facts and data and ideas. During the much more typical periods of slow scientific development, it is usually hard to see identifiable progress. One just can't tell which of the numerous experiments being performed will turn out to be important. Sometimes the key to unlock new doors is not a discovery of a new phenomenon but a more accurate measurement of an old one. One of the least glamorous, but most important, aspects of scientific inquiry is the quest for precision, the drive to increase the accuracy of previous measurements "to the next decimal place." This

[8] Support of astronomy has often been considered a sign of madness. In 1926 the heirs of William Johnson McDonald were startled to learn that he left the bulk of his fortune, over $1 million (a tidy sum in 1926), to the University of Texas to construct an observatory. The heirs contested the will, arguing that the fact that he gave money for astronomy was prima facie evidence of insanity. After years of litigation the matter was settled out of court and the McDonald Observatory became one of the world's foremost centers of astronomy.

was Tycho's greatest skill. He did not revolutionize astronomy by a single great discovery. His most important contribution to astronomy was a body of precision measurements that were the result of a lifelong program of meticulous observations.

The heavens seemed to speak to Tycho. Between the solar eclipse that led him to become an astronomer and the supernova of 1572 that established him as the leading astronomer of his time, a third celestial event shaped his life and work. On the night of August 24, 1563, a conjunction of Saturn and Jupiter occurred, during which the planets appeared so close to each other as to be indistinguishable. Just as with the solar eclipse three years earlier, the conjunction was predicted. But the prediction was wrong, or at least not precise enough to suit Tycho. The old astronomical tables based on the Earth-centered solar system of the **Ptolemaic system** were in error by a month in predicting the date of the conjunction. Even the latest prediction based on the Sun-centered **Copernican system** was off by three days. This inconsistency greatly troubled Tycho. He knew that something was not right with the cosmological models of the day, and he dedicated his life to finding the correct arrangement and motions of the planets. Although he would not see the final resolution, his patient observations would provide the

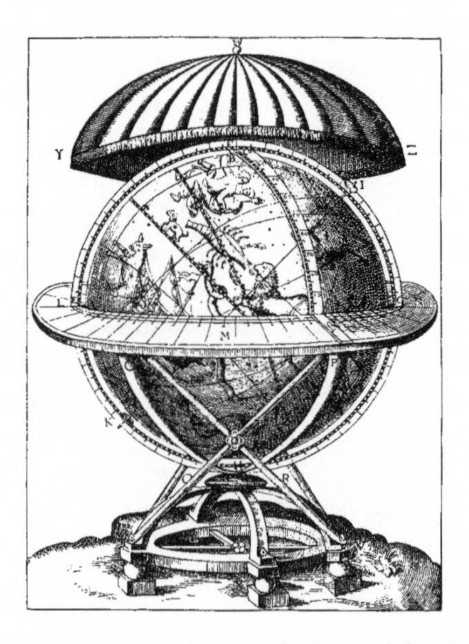

Tycho recorded accurate star positions on the brass globe pictured here. It was over thirteen feet in diameter.

canvas on which his successor, Johannes Kepler, would complete the planetary picture.

Tycho realized that before the correct theory could be developed, astronomical observations of much greater accuracy than those available must be performed. Some measurements in science are done to test a theory, and some are done to accumulate sufficient data to develop a theory. What Tycho set about to do at Uraniborg was to "survey the sky," making the most accurate and the most complete observations possible.

Once Tycho had accumulated a database of observations, he set about constructing a model of the universe that would serve as a new cosmology, a cosmology that could be combined with his observations to describe accurately the motions of the stars and the planets.

In Tycho's education he would have learned of two cosmological models. The first model (p. 28) was a geocentric model, with Earth at the center of the solar system. In this model the Sun, Moon, stars, and planets all revolve about the stationary Earth. This model was inspired by the cosmological worldview of Aristotle, and was perfected in the second century of the Christian Era by the Greek astronomer Ptolemy. In order of increasing distance from the center of the solar system (Earth) were the Moon, Venus, Mercury, the Sun, Mars, Jupiter, and Saturn (the remaining planets—Uranus, Neptune, and Pluto—were not yet discovered because they cannot be seen with the unaided eye). Even without the added complexity of epicycles, eccentrics, deferents, and equants necessary to account for the observed motion of the planets, the arrangement is incorrect.[9] Even if we remove the circles describing the orbits and ignore the question of the center and which object is at rest, the arrangement is still wrong.

The second model (p. 29) on the market in Tycho's era was a heliocentric model with the Sun at the center of the solar system, and Earth rotating on its axis and orbiting the Sun along with the other planets. Although this model had been discussed in antiquity, most notably by the Greek astronomer Aristarchus in the second century B.C., it had been most recently championed in 1543 by the Polish astronomer Copernicus. Although this model is also incorrect because it assumed the planets trace out perfect circles and move with constant speed (and

[9] Epicycles, eccentrics, and so on are defined in the discussion of the Ptolemaic model in "The Devil in the Details" section at the end of this book.

also had more than its share of epicycles, eccentrics, deferents, and complexities), Copernicus did have the correct arrangement of the solar system: the Sun at the center, with Mercury, Venus, Earth, Mars, Jupiter, and Saturn orbiting at increasing distances about the Sun, with the Moon orbiting about the Earth.

To say that these are "competing" models would not be fair. Almost all educated people of the time believed in the Ptolemaic model (to the extent that they "believed" in any model). The grip of Aristotelian physics and cosmology in the early Renaissance was so great that Copernicus's theory was not taken seriously by many people. The hegemony of Aristotle in scientific matters was not to crumble until after the onslaught of Galileo in the 1630s. The scientific procedure of Aristotle consisted of an attempt to explain nature by "discovering" the causes of natural phenomena. Modern cosmology and astrophysics also seeks an explanation of celestial phenomena in terms of their causes. Today the causes are taken to be the laws of physics, as determined by terrestrial experiment. The crucial difference is that for Aristotle, the causes were determined by sense or by philosophy, rather

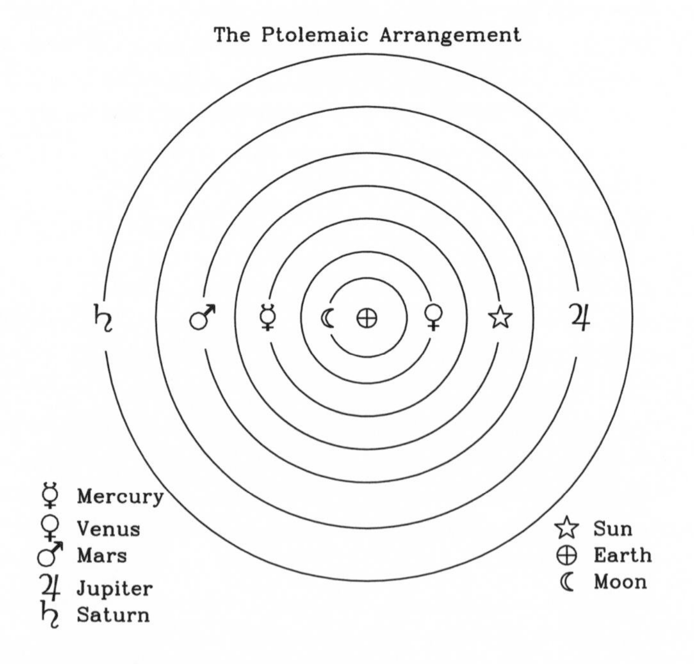

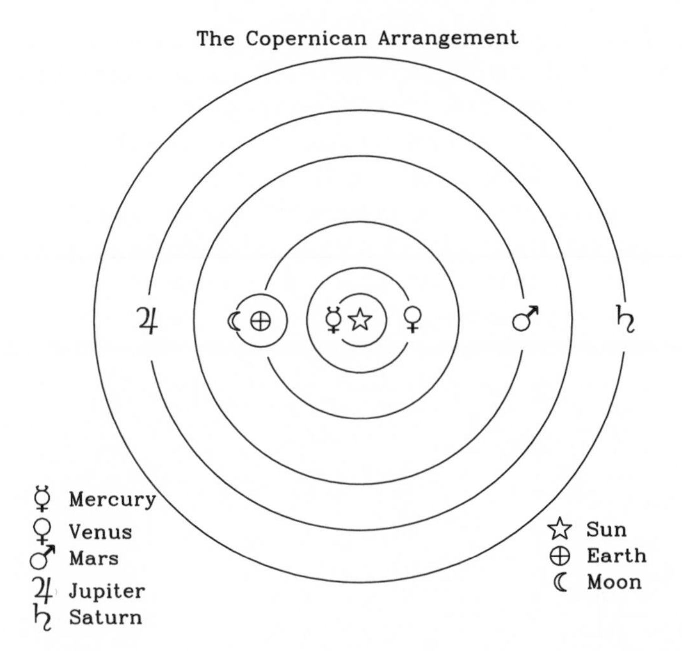

than by experience. In modern physics, experiment is the ultimate arbitrator, not philosophy.

Aristotle's cosmology is presented in his book *De Caelo.* For philosophical reasons, Earth is taken to be at rest, in the center of the universe. The stars, the Moon, the Sun, and the planets all circle about Earth at a fixed distance. In the time of Tycho it was believed that the planets traveled on crystalline spheres, propelled by mechanical gears. Since the heavens are beyond the corrupting influence of humans, the motion of celestial bodies must be perfect, and it was thought that the perfect geometric figure is the circle. Furthermore, the planets must always travel with the same speed. Perfectly circular motion with constant velocity about a fixed center is known as *uniform circular motion.*

After Aristotle's death, in the century before the Christian Era, Greek astronomers realized the original scheme of uniform circular motion did not fit observations. The Greek astronomer Hipparchus came to the conclusion that either Earth moved or the motion of the planets, Sun, and Moon was more complicated than the simple uniform circular motion of Aristotle. Another Greek astronomer, Geminus, proposed a compromise solution to Hipparchus.

Although conferences would not flourish until scientists invented funding agencies, one can easily imagine a setting where Hipparchus presents the summary talk titled "Problems in Reconciling Aristotelian Cosmology with Recent Astronomical Observations" at the 227th Annual International Cosmology Conference, held at the Athens Holiday Inn, June 2–14, 114 B.C. After Hipparchus's talk, the chair of the session asks for questions. At first there are no questions—most of the audience is suffering from shiplag and some are sleeping. But a young, unknown assistant professor, Geminus, wants to impress people so he can get tenure. So he boldly stands up and says:

> I don't see why the most important problem in astronomy should be to reconcile the standard model of Aristotelian physics with astronomical observations. Look guys, we have no clue as to the causes of the (apparent) motions of heavenly bodies. Our understanding of motion, inertia, and forces is too primitive. Perhaps there is something wrong with Aristotelian physics, perhaps the physics of earthly bodies does not apply to heavenly ones. Perhaps Aristotle's logic and physics are right but we just aren't smart enough to apply them correctly. In any event, let's just forget about the causes. Let's abandon (what will eventually come to be known as) physics as a basis for astronomy for a short time until the aforementioned difficulties are solved; then we can rejoin them. We will temporarily stop work on "physical" astronomy, and start work on "mathematical" astronomy. Henceforth, we will regard our models of eccentric spheres, epicycles, this going around that, and so on as mere mathematical illusions. Of course, we all know that the truth of the matter is that Earth is at rest at the center of the universe, and actual celestial motions are perfect uniform circular motions according to the philosophy of Plato and Aristotle. Let's not worry so much that this is not what is observed. Since we don't understand the causes of the motions, the job of an astronomer should be to make astronomical observations in order to construct mathematical models of the motions so that we may "save the appearances."[10]

[10] The Greek word *phenomena* means "appearances." "Saving the phenomena" means reconciling what is believed to be true with what is actually observed.

Everyone thinks this is a great idea. Another Greek cosmologist, named Dimitrius,[11] older and bolder, stands up and says that it would be just a temporary separation between astronomy and physics, because he is very close to finding the ultimate "theory of everything." Sadly, as usual, poor Dimitrius is wrong. Another eighteen hundred years would pass before Newton put astronomy and physics back together in 1686. After the usual criticism of Dimitrius, the session adjourned, and everyone enjoyed a fine afternoon at the beach and prepared for the conference banquet. The after-dinner speech at the banquet was given by the ancient and wise natural philosopher, Leonus.[12] After a few familiar jokes and a plea for greater support of science from the Athenian assembly, Leonus embraced the proposal of Geminus as the best approach to reconcile observation and theory.

The proposal of Geminus was a reasonable, perhaps inspired, response to the situation in 114 B.C. But the aftereffects of the messy divorce of physics from astronomy were felt long after the reasons for the divorce were forgotten. In the universities of the Renaissance, physics and cosmology were taught by professors of natural philosophy, whereas astronomy was taught by professors of mathematics. The cosmological model of the universe was considered a philosophical issue. Physics described the causes of the heavenly motions, whereas astronomy merely developed mathematical models to save the phenomena. In a typical academic turf battle, natural philosophers believed cosmology to be part of their bailiwick since they were granted custody in the divorce, and they regarded the work of astronomers as mathematical exercises unrelated to the actual arrangements of the heavens.

Although many astronomers could comfortably accommodate themselves to work as phenomena savers, a restless few wanted more. Certainly Copernicus had not labored a lifetime to develop a model merely to save the appearances. He believed that Earth moved! Tycho also thought that he could do more than construct a model. He wanted to determine the actual arrangements and motions of the heavenly bodies.

Tycho never seriously entertained the possibility that Earth moved, describing it as a "hulking, lazy body, unfit for motion." He simply dismissed the idea as absurd. He did see a simplicity in the arrangement of Copernicus, however, and he did realize that the stationary-Earth model

[11] This is a fictional character.

[12] This is a fictional character.

of Ptolemy simply didn't work. Tycho set about constructing a model of the solar system, as he put it, "expounding the motions of the planets according to the models and parameters of Copernicus, but reducing everything to the stability of the Earth." In this way he could avoid "both the mathematical absurdity of Ptolemy and the physical absurdity of Copernicus."

Tycho developed his own hybrid model. In his stationary-Earth model, the Sun and Moon orbit Earth and the other planets orbit the Sun. A glance at Tycho's arrangement reveals two striking features. First, if one removes the circles indicating the orbits of the planets, it is exactly the arrangement of the solar system according to Copernicus. In fact, if one starts with the Copernican model and simply assigns the state of rest to Earth rather than to the Sun, one immediately obtains Tycho's model. For Tycho to start with Copernicus's model and obtain his model in this single step of transferring the state of rest would have been a completely trivial task. Even in 1580 it would have been the work of an afternoon. But this is not the path Tycho took. He arrived at his model by a tortuous route taking several years of incredibly hard work. He

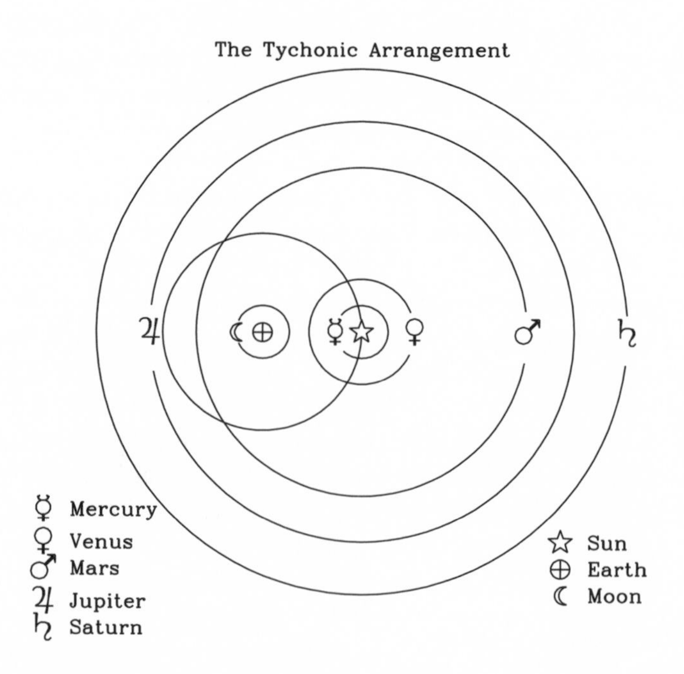

started with the model he regarded as essentially correct, the Ptolemaic model, and started modifying it to agree better with the results of his years of precise observations. It seems that Tycho never did realize, certainly never did admit, that he had simply rediscovered the Copernican arrangements of the heavens, but with a different body at rest. Clearly Tycho's path from point *A* to point *B* went through point *Z* because he was steered by his model of a stationary Earth.

The second, and most curious, feature of the Tychonic model is that the orbit of the Sun intersects the orbit of Mars (as well as the orbits of Mercury and Venus). Until this time, Tycho, as well as his contemporaries, believed that some physical entity existed in space to carry the heavenly bodies through their appointed rounds. Here Tycho required no small amount of courage to take his model seriously. Try as he might, he couldn't have the arrangement of Copernicus and a stationary Earth without smashing the celestial spheres, part of the very machinery believed necessary to run the universe. A more timid soul would have been afraid to smash the spheres, but Tycho knew that there was already a crack in the crystal.

Some years before, on the evening of November 13, 1577, Tycho noticed another new "star" in the western sky. Within a few hours he was able to discern a tail on the star, and he immediately realized that he was observing a comet. Tycho had been patiently awaiting a comet to study since the new star of 1572. In his book on the new star, he predicted that comets were also "above" the Moon (farther away from Earth than the Moon). Finally he had the opportunity to study the path of a comet for himself.

To determine the distance to the comet Tycho had to search for an apparent **parallax** of the comet. Determining the parallax of the comet was complicated, because unlike the new star of 1572, the comet was obviously moving across the sky with some intrinsic motion of its own.

For two months he performed painstaking observations of the positions of the comet, until finally on January 26, 1578, it faded from sight. Tycho was unable to determine a parallax for the comet, which led him to conclude that the comet had to be more than three hundred Earth radii away. According to the scales of the solar system in use at the time, this placed the comet outside the orbit of Venus. Tycho was overjoyed (nothing is more enjoyable than making a prediction and being right). Tycho also noticed that the tail of the comet always pointed away

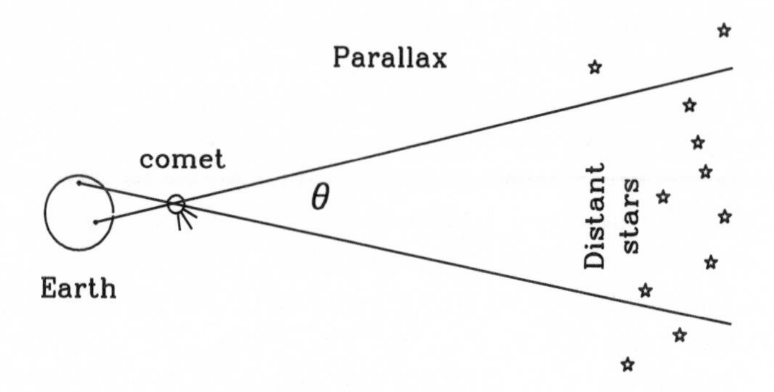

Parallax and the distance to the comet: If the comet was near Earth, then observers in different locations (or because of the rotation of Earth at the same location at different times) would see the comet in a different apparent location against the background of stars. If the angle θ is measured, and the distance between the two observers is known, then the distance to the comet can be determined. The more distant the comet, the smaller the angle θ.

from the Sun. Together with other observations, this led him to conclude that the comet orbited the Sun.

Tycho's conclusion that the orbit of the comet was more distant than the Moon had two tremendously important ramifications. Again, something in the ethereal part of the universe had changed, in contradiction to Aristotle's philosophy. Equally troubling to the philosophers was that if the comet started beyond the orbit of Venus, either it mysteriously had to pass undisturbed through the crystalline spheres, or the spheres didn't exist.

Again, the blind watchers of the sky came to the opposite conclusion in the more than two hundred manuscripts, pamphlets, and books written about the comet. Of course, many of them "knew" beforehand from the teaching of Aristotle that the comet was an atmospheric phenomenon. If it was atmospheric, they should be able to measure a parallax for the comet . . . and they did. They made two crucial mistakes in their observations. First, they used the standard method of the great astronomer and mathematician Regiomontanus (1436–1476) to determine the parallax. Tycho realized that the method of Regiomontanus could be used to determine parallax only if three conditions were met: (1) the object did not have any intrinsic motion of its own, (2) there was no uncertainty in the timing of the observations, and (3) the object could be observed on the meridian. None of these conditions pertained to the comet. Tycho realized this and developed his own methods for determining the parallax. The second error made by other astronomers was the use of the star tables of Ptolemy or Copernicus for reference positions. Tycho knew that those tables were unreliable. Over the years in his survey of the sky he had developed much more accurate positions for reference stars. Again the search for seemingly needless precision pays off! Tycho was simply a better astronomer than his contemporaries.

✦ ✦ ✦

Tycho had cracked the crystalline spheres in his study of the orbit of the comet of 1577, and he was prepared to smash the crystalline spheres with the Tychonic model of the universe. It was a small price to pay, for Tycho thought he had found the arrangement of the heavens.

Galileo and Kepler knew that the arrangements of Ptolemy, Copernicus, and Tycho were actually mathematically identical. It is not impossible that deep inside Tycho realized this also. If one is simply interested in a cosmological model to predict the occurrence of eclipses or to cast horoscopes, the Ptolemaic model is as good as any. In fact, if today you open any encyclopedia to the section on solar eclipses, you will find a diagram showing Earth at the center of the solar system. Any astronomer who wants to predict when a solar eclipse will occur will do the calculation in a "reference frame" in which Earth is at rest and the Sun travels around the zodiac.

But in science, one demands more of a model than simply the ability to predict the result of observations. Models have to be more than a simple mathematical tool. One demands of a model that it lead to deeper truth. The reason the heliocentric model of the solar system is better than the geocentric model isn't that the calculations are easier or more accurate but that it allows the reconnection of physics with astronomy. It is the model of Copernicus, modified by Kepler, which Newton used to explain the motions of the Moon and the planets on the basis of physical law. It is this model that is encompassed by Einstein's general theory of relativity. The heliocentric model is the fertile one. The Ptolemaic model and the Tychonic model are worse than wrong; they are barren.

But most natural philosophers and mathematical astronomers of the day did not take seriously the question of which of the three models was the best cosmological system. To them it was a nonissue. They could not free themselves from the idea that astronomy was simply a mathematical exercise to save the phenomena. It wasn't so much that they clung to the wrong model; they didn't see a real advantage to any of the models. The agreement that astronomy was separate from physics had been made so long in the past that it had been forgotten that deep down inside was still the goal of explaining the phenomena, not merely saving the phenomena.

Tycho's cosmology offended everyone. He demanded an acceptance of Aristotelian physics by astronomers. He forced astronomers to accept that Earth was immobile at the center of the universe. He also wanted the natural philosophers to alter their arrangement of the universe and accept the notion of other planets orbiting the Sun.

Most academics of the time ignored Tycho's model. Again Tycho railed against the blind watchers of the sky. In this instance, however, Tycho shared in the blindness. His model was wrong. Although it was wrong, this step in the wrong direction may have been necessary before the correct model could be developed. Tycho's smashing of the celestial spheres may have been a prerequisite for Kepler to take the next step and say that perhaps the orbits of the planets were not circles after all.

TYCHO IN EXILE

By 1588, Tycho Brahe had established in his twelve years on the island Hveen the world's first "national laboratory," the first "big science" effort the world had ever seen. But in the spring of 1588, at the high-water mark of Tycho's scientific work, Fredrick II, Tycho's patron, died. Fredrick's son and heir, Christian IV, was then a child of twelve. Until Christian reached the age of twenty, the country was ruled by a protectorate appointed by the Rigsraad. The protectorate was not so kindly disposed to spending vast sums of state revenue for mere astronomy. It soon became clear that the days of Tycho's royal patronage were limited.

It is often the case that the favor of scientific projects changes with administrations. Tycho should have seen the writing on the wall in his dealings with the regents in the years of Christian's minority. It was, and still is, the common way of life for scientists to trim their sails in response to the political winds. In times of political or social hostility, scientists bide their time and wait for a more responsive administration. Tycho was a Great Dane, however, who would not bow to any man and would not beg from any king. His arrogance was his undoing. He dared to imply in correspondence with the King that his scientific accomplishments made him the equal of the twenty-year-old king.

The young king was furious. How dare a mere astronomer speak in such a way to royalty. In a remarkably frank and straightforward correspondence, Christian wrote to Tycho, stating in no uncertain terms just how he valued the royal astronomer and mathematician:

> ...[how dare you] not blush to act as if you were my equal....I expect from this day to be respected by you in a different manner if you are to find me a gracious lord and king....Your correspondence is somewhat peculiarly styled and not without great audacity and want of sense, as if I should account to you why and for what reason I made any change about [your share of] crown estates....I graciously answer that if you [are to] serve as a mathematician and do what you wish to do, then you should first humbly offer your service and ask about it as a servant ought to do....When that is done, I shall afterwards know how to declare my will.

History is rife with episodes of genius mistreated by employers or rulers who seemed not to value their services. One thinks of Michelangelo bullied by Pope Julius II, or Mozart forced to dine with the other "servants" of Emperor Joseph II. As much as they may have privately chafed under such treatment, a Mozart or a Michelangelo was not in a position to do anything about the humiliation that went with royal or papal patronage. Tycho was. He poured out his soul in a poignant *Elegy to Denmark*, gathered together his resources, left his island and his country, and lived the last years of his life in self-imposed exile. He wandered about Europe (have quadrant, will travel) for two years before settling in Prague. He wrote during his travels:

> An astronomer must be cosmopolitan, because ignorant statesmen cannot be trusted to value their services.

But the view of Tycho as mistreated genius is not the complete picture; there was another side to him. Tycho was not a gentle genius, but an imperious taskmaster in his own right. He regarded the other inhabitants of his island as serfs, to be punished or tortured if they dared to complain about taxes. His fellow islanders certainly did not see him as mistreated. Perhaps a third view of Tycho was that of someone named Jepp. As an astronomer king Tycho had a court of sorts, and Jepp was his buffoon. Jepp was a person who stood up and looked Tycho squarely in the knees—a dwarf whom Tycho treated more as a pet than as a person. During dinners at Uraniborg, Jepp would sit under the table at the feet of the astronomer king begging for food, and occasionally Tycho would toss a scrap of food under the table for his buffoon. Although this behavior toward another human being is contemptible, it is impossible to make a moral judgment of people of another time on the basis of modern standards. It is perhaps too much to expect that because Tycho was advanced in his scientific outlook he should be equally advanced in his social or moral outlook.

The spectacle of Tycho wandering about the continent with his entourage of servants, his instruments, and his dwarf, seeking employ-

Tycho and assistants observing with the great mural quadrant. Just like the pharaohs of Egypt, Tycho is pictured towering over the minuscule assistants. Only Tycho's loyal hunting dog, obviously of higher station than Tycho's assistants, is shown in proper proportion.

ment, was not the only squalid episode of the later part of his life. One of the ugliest aspects of science is arguments over priority of scientific discoveries. Tycho clearly was not the first, and not the only, person to imagine the arrangement of the heavens according to what we know as the Tychonic model. Another astronomer, named Ursus, also claimed the model as his. Ursus was as different from Tycho as one can imagine. Ursus was born into abject poverty, working his way through school as a swineherd. In 1584 he had worked for Tycho on Uraniborg. One day he was discovered snooping around Tycho's office, and for this indiscretion he was thrown off the island. Ursus eventually landed on his feet and became imperial mathematician to Rudolph II of Bohemia, the post that Tycho himself would later fill.

In 1584, four years before the Tychonic system would appear in print under Tycho's name, Ursus published a book proposing the same basic model. Ursus's model was not as developed or as polished as Tycho's, and also had the fundamental difference that the orbits of Mars and the Sun did not intersect. Tycho had been playing with this version of the model when Ursus was on Hveen, but realized that this model would result in predictions in gross disagreement with the observed motions of Mars. Tycho knew that the price of a stationary Earth was a smashing of celestial spheres.

Tycho was furious when he saw Ursus's book. He was certain that the model of a Brahe, a leading family of Denmark, had been stolen by Ursus, a former leader of pigs, during one of his surreptitious forays into the inner sanctum of Tycho's castle. Tycho attacked Ursus mercilessly, using all of his tremendous influence to destroy him. Ursus fought back as best he could, but he did not rise above the pettiness of claiming that Tycho had made observations by sighting stars through the holes at the end of his golden nose. Many friends advised Tycho to drop the unseemly matter. He received advice from an astronomer Rollenhagen, his old and trusted friend: "This I surely know, if I wrestle with dung, win or lose, I am always defiled." Although Tycho succeeded in destroying Ursus, he was indeed defiled in the process. Sadly, the lesson from all of this has never been learned, and such ugly incidents are still a common part of life today.

Tycho finally landed an acceptable position as mathematician and astronomer to Rudolph II, and he began arrangements to build another, even grander observatory. But big science projects require not only money but also trained and talented people. To staff the new observatory, Tycho had to replace the assistants he left behind in Denmark. He wrote disparagingly of the dim-witted assistants he had with him, and set about to recruit the best astronomers in Europe. Among those was one Johannes Kepler, who would succeed him and set the heavens in their proper place.

But Tycho did not survive to see the fruits of his labors. In 1601, barely a year after starting a new life in Prague, he attended a dinner as part of his busy social life. Kepler records the unfortunate events of October 13, 1601:

> Tycho accompanied Royal Councillor Minckwicz to supper at the home of Peter Rozmberk. Holding his waters longer than he was accustomed, Brahe remained seated. Although he drank a bit overgenerously and felt pressure on his bladder, he had less concern for his health than manners, and remained at the table. By the time he returned home, he could no longer urinate.

The price Tycho paid for etiquette was acute uremia.

Kepler also chronicled the last hours of the great Tycho. For eleven days Tycho drifted in and out of consciousness, delirious during his waking moments. Kepler notes that during the final night of his life Tycho kept whispering, "Let me not seem to have died in vain." The astronomer king, who would bow to no man and beg from no king, was now bowing to inevitable death and begging Kepler to continue work on the Tychonic system of the universe. This is all the sadder because Tycho surely must have known that Kepler would do no such thing, that Kepler was a confirmed Copernican. Perhaps Tycho himself knew that his model of the universe would not survive. After a lifetime of meticulous observations to lay the groundwork for a new world system, Tycho knew he would not even be allowed to glimpse the promised land.

Tycho was buried on November 4, 1601.[13] One of greatest Danes of all was buried in the cathedral of Prague following ceremonies befitting an astronomer king. Tycho's statue still may be found in Prague's Teyn Cathedral, with the inscription that seems to be addressed to the heavens: "To be, rather than to be perceived."

Although Tycho had not lived to see his (incorrect) model of the universe supplant Ptolemy's, he did bequeath to astronomy the most

[13] A final sad irony is that the greatest astronomer in the thirteen hundred years since Ptolemy died at age fifty-five, a mere six years before the invention of the telescope.

complete and accurate observations of the heavens ever performed. His data were not available to astronomers by the time of his death, however. Tycho had always been a secretive person, and the episode with Ursus drove him to even greater secrecy. At the time of his death much of his data was unpublished, and had not even been seen by his assistants. Kepler desperately wanted the data, but someone else did also.

Although he showed no great ability or originality, Franz Tengnagel, one of Tycho's dim-witted assistants, came up with a plan to advance himself above Tycho's other students and colleagues. It was an ingenious plan, an inspired plan, a totally original plan: he would marry the boss's daughter. Late in 1600 he summoned up his courage, went to Tycho, and said "I want to marry your daughter." Tycho was furious and threw poor Tengnagel out on his ear. "No daughter of mine is going to marry an astronomer," Tycho might have replied, "at least not a second-rate one." Tengnagel might not have been a talented astronomer, but he was a clever young man. In fact, Tycho greatly valued his services as a courier and emissary. A few months later Tengnagel went to Tycho again, but took a completely different approach. This time he said to Tycho, "I want to marry your daughter—*she's pregnant*." Elizabeth Brahe and Franz Tengnagel were married in the summer of 1601, and presented Tycho Brahe with a grandchild shortly thereafter.

After Tycho's unexpected death, Tengnagel took possession of Tycho's instruments and prepared to assume the role of Tycho's successor, keeping the business in the family as it were. Before Tengnagel could secure them, however, Kepler stole Tycho's observational logbooks. Kepler desperately needed Tycho's data to complete his studies of the orbit of Mars. Although he had no legal rights to the logs, he had no intention of surrendering them to Tengnagel. Kepler realized that Tycho's observations were a treasure that would allow him to unlock the secrets of the planetary motions. Tengnagel could never make use of the data, but he could legally prevent Kepler from publishing results based on the stolen logbooks.

Tengnagel proposed a deal to clear the impasse. He would allow Kepler to use Tycho's data, with the stipulation that he would appear as an equal coauthor on any book Kepler would publish. Tengnagel wished to be guaranteed equal credit in all Kepler's discoveries. In other words, Tengnagel was attempting nothing less than immortality by blackmail! By marrying the boss's daughter, he would ensure that his name would be part of the three basic laws of planetary motion. The only thing more shocking than Tengnagel's sheer audacity is Kepler's response. He readily agreed! Kepler said that he would be happy to have Tengnagel's name on his book if, in addition, Tengnagel would grant to Kepler one-quarter of his annual stipend from the government.

So here we have the rare opportunity to find a quantitative answer to the often asked question "What is the price of immortality?" It may surprise you that such questions can be answered. For instance, did you realize that the question "What is the value of a human life?" can be easily answered? To find a value of a human life, ask a Mafia hit man. It is $3,500 in Chicago; $3,200 in Detroit; and even less in Cleveland. The price of immortality is smaller still. Apparently an upper limit to the price of immortality is something less than $250: Tengnagel refused Kepler's offer. In the end, Tengnagel realized he couldn't do much with the data, so he gave them to Kepler with the stipulation that he be allowed to write a preface to his next book. Kepler gladly accepted. Tengnagel's pompous preface can be found in Kepler's *New Astronomy.*

No one devotes a lifetime to science in the hope of making small advances. Every scientist secretly or overtly hopes to make great discoveries. Certainly the proud astronomer king of Denmark was no exception. Tycho knew that the prize could not be higher. As the French mathematician Joseph-Louis Lagrange wrote, "it is given to only one person to discover the system of the world." Tycho Brahe wanted to be that person. As Stephen Hawking put it, the goal of a cosmologist is "to know the mind of God." Tycho Brahe wanted that knowledge. He was denied. He went to his death believing that the Tychonic model was his lasting contribution to astronomy. The astronomer who had seen farther than any previous astronomer in the end died a blind watcher of the sky.

THREE

Wars of the Worlds

What sort of early-childhood environment is conducive to the development of genius? Does the environment play any role at all? Are people born with the indelible mark of genius stamped on their foreheads, seared on their brains, carved in their hearts, or does it develop from having their curiosity piqued by a magical experience at some early age? Is there something anxious parents can do to ensure their little child's intellectual development? If for some reason you want to be the parent of a genius, should you follow recent advice columns and fill the nursery with the music of Mozart, paper the walls with pretty colored pictures, or play French conversation tapes during feeding times? What's the prescription for the magic genius pill to give to children?

Perhaps we can make some headway on the nature-nurture question by doing what an astronomer would: performing an observation. Let's journey in time and space to the year 1575, to an area of southwest Germany cradled between the rivers Neckar and Rhine, just on the outskirts of the Black Forest near the city of Stuttgart. In this idyllic corner of the world we find the picturesque city of Weil der Stadt. At the end of a typical winding street of the city we see a small house. In one of the three small rooms of the house is the subject of our study, a four-year-old boy named Johannes Kepler. As we walk down the street toward the house we have an uneasy feeling that in another four hundred years or so there will be train tracks somewhere near, and we have just crossed to the wrong side of them. We won't enter the house, because we don't want to disturb the subject of our study. Luckily, one of the small, unpainted shutters on the house seems to have fallen off. Let's take a quick, unintrusive peek through the bare window and look around Kepler's childhood home. This is a rare opportunity to examine the surroundings of a child destined to become a towering scientific genius. We view the people and the house clearly over the 420-year time span through the chronicles in which Kepler recorded memories of his childhood. A patient and accurate watcher of the sky, Kepler also viewed his family, as well as himself, with the same dispassionate eyes. Probably no other scientific genius of Kepler's rank left behind such an extensive written record of himself and his environment. What will we find in the boyhood home of the nascent genius? Will we find music, learning, literature? How about at the very least some pretty pictures on the wall of his nursery?

Let's first look for his dear parents. Well, we won't find them at home. Kepler has just been abandoned by his parents, probably to his advantage. His father is "vicious, inflexible, quarrelsome, and a wanderer; beat his wife often." (The quotes are from Kepler's chronicles.) Dear old dad is a military adventurer. This year's adventure is to fight the Protestant insurgents in the Netherlands. It seems unlikely this action is the result of religious fervor, since the Keplers are Protestants themselves. No, Dad bounces from army to army to navy, fighting wherever the pay is good and adventure high. His only redeeming virtue seems to be that at least he is honest enough to admit that he kills for fun and money, and doesn't claim to kill in the name of God, king, or country. He passes through Weil der Stadt every so often whenever there is peace in Europe, which is to say rarely, to beat and bully his children. He was last seen in the pages of history in 1588, sailing off with the Neapolitan fleet. How about Mom? The four-year-old Johannes might cry for his mother (not that he would be heard above the din of the crowded household), but Mom enjoys life as a camp follower, and is with Dad in the Netherlands. What does Kepler miss with Mom not around? Kepler's mother, an innkeeper's daughter, is "small, thin, swarthy, gossiping, and quarrelsome, of a bad disposition." She was raised by her aunt, who was burned at the stake as a witch. Kepler's mother has developed quite a reputation of her own for witchcraft. She is feared by everyone in town because of her knowledge of herbs and magic, and later escaped her aunt's fate only through the intervention of her son Johannes.

The head of the household is Kepler's maternal grandfather, Sebald. He is "remarkably arrogant, short tempered and obstinate—his face betrays his licentious past." He has a red and fleshy face; obviously, he is fond of good German beer. In 1575, when we see this snapshot of Kepler's youth, Grandfather Sebald is seventy-five years old and in full mental free fall. His wife, Katherine, Kepler's grandmother, is "restless and clever . . . a liar; of a fiery nature, an inveterate troublemaker, jealous, extreme in her hatreds, violent, and a bearer of grudges."

Of Kepler's three siblings who survived childhood, the closest in age to him was his brother Heinrich. Heinrich was also nearest to Johannes in personality. Heinrich was constantly ill. He was an epileptic and seemed to always be near death from disease, accidents, or beatings. He left home to avoid being sold by his father, but returned penniless and died unmarried and unloved.

Also sharing the small house in Weil der Stadt were a dozen aunts, uncles, and cousins, all in the family mold. A couple of examples will complete the picture: Uncle Sedaldus was, at various times, "an astrologer, a Jesuit, acquired a wife, contracted the French sickness, was vicious and disliked by fellow townspeople." Aunt Kunigund is not around either; she had died the year before—poisoned they said (no shortage of suspects in that household). Another aunt "lived sumptuously, squandered her goods, and is now a beggar." The list of miscreants goes on and on. Here is a household that gives new meaning to the phrase "dysfunctional family."

What of the four-year-old Johannes Kepler? There, sitting in the corner observing the bedlam of his home is a puny, sickly, unattractive child. There sits a child destined to confront the blind watchers of the sky, destined to look into the heavens and to see farther than anyone before, squinting blankly into the distance. The person who will succeed Tycho Brahe as imperial mathematician and court astronomer to Rudolph II of Bohemia is comically nearsighted and suffers from multiple vision since birth. Nowhere is there better evidence that the clarity with which we see the sky is unrelated to visual acuity. Perhaps the one kindness bestowed on young Kepler is the myopia that prevents him from seeing even more clearly the people in his immediate gene pool. Can we find a less likely candidate in a less likely environment to take up the mantle of European astronomy and prepare the way for Isaac Newton?

A prudent observer would quietly replace the shutter to the Kepler home and slink away before being seen and attacked, or, scarier still, being seen and invited to enter the household. An observer concerned with the common good of the town might wish to board up the windows and doors lest some family member escape and reproduce further (reproduction seemed to have been the only family talent). A soft-hearted observer might wish to help the family. But whom should be called first: the police, the welfare agency, an army of social workers, a minister, a doctor, a lawyer, or perhaps a fight referee?

What of the nature-nurture observers? What do they take away from this experience? How will the naturists explain that from a father who narrowly escaped the gallows, and from a mother who equally narrowly escaped the stake, would issue the founder of modern celestial mechanics? How will they find it possible to imagine that the myopic child suffering from multiple vision would become the father of the science of modern optics? How will the nurturists explain that a genius of the first rank would emerge from the squalor of the Kepler house?

As is usually the case, both sides would claim victory. The naturists would claim that Kepler was simply born a genius, destined for greatness, and not even the environment in which he grew up could change it. The nurturists would say that without the childhood that Kepler experienced, he would not have developed the neurotic, insecure personality that seemed very much a part of his unique brand of genius and drove him to accomplish all that he did. The answer, of course, is that there is no answer. The nature-nurture question is not a scientific one. No experiment can be performed that could falsify either theory. "Not science" doesn't imply not interesting or not important; perhaps it just means not answerable.

What possible advantage did this little urchin have? In a strange way, he benefited from the turmoil of religious conflicts and disputes sweeping Europe at the end of the sixteenth century. Although he would constantly change jobs, trying to stay one step ahead of religious wars and revolutions, and eventually he would be swept away by the tides of the Thirty Years' War, religious conflicts were responsible for strengthening the system of higher education throughout the continent of Europe. All sides realized that battles for the souls of the masses were fought from the pulpit, as well as from behind canons. It was crucial to have a literate, educated clergy in the battle, and the training grounds for the clergy were the universities. The Lutherans controlling southern Germany realized this and developed an educational system second to none, so Kepler had excellent public educational opportunities in his native land.

It seems somewhat of a miracle that Kepler went to school at all. During his early years schooling was often interrupted. From the age of nine to eleven, he was taken out of school and put to work in the country. Not until age thirteen was he able to attend school with a purpose. His purpose: to become a Lutheran minister.

He was no happier in school than he was in his grandfather's home. Kepler was a hopeless hypochondriac, with complaints of imagined illnesses interrupted only by real sickness. In his chronicles he writes:

> At the age of four I nearly died of smallpox.... My hands were badly crippled....During the age of 14–15, I suffered continually from skin ailments, severe sores, scabs, putrid wounds on my feet which wouldn't heal and kept breaking out again. On the middle finger of my right hand I had a worm, a huge sore on the left hand....When 16 I nearly died of a fever....When 19 I suffered terribly from headaches and disturbances of my limbs....I continually suffered from the mange and the dry disease....At the age of 20 I suffered a disturbance of the body and mind....

Kepler complained of terrible hemorrhoids, which required him to work standing rather than sitting at a table. He sums up his health: "I think I am one of those people whose gall bladder has a direct opening into the stomach; such people are short lived as a rule." Kepler's chronicles read as a classic case of hypochondria. He complained that he did absolutely everything in his power to heal himself, even going so far as to bathe once in his lifetime. He did this against his better judgment after constant nagging from his wife.[1]

Physical ailments were not his only problem. It should not surprise anyone to learn that the sickly child from the brutal home life in Weil der Stadt was not popular at school. He was continually beaten by his classmates—Kepler writes that they were jealous of his intellectual

[1] The bath nearly killed him: "its heat affected me and constricted my bowels," he reported.

abilities. His remembrances of his school days are as depressing as the tales of his physical ailments:

> I hated Kolinus.... Braunbaum was my enemy.... I willingly incurred the hatred of Seiffer.... Ortholphus hated me as I hated Kolinus.... Kleberus hated me as a rival.... My talent made Rebstock hate me.... Husalius opposed my progress.... Jaeger betrayed me....

He goes on to list some twenty-two enemies in his chronicles. Hypochondria mixed with acute paranoia does not produce a happy youth.

It should also not surprise anyone that the tortured youth was not lucky in love. Again, from his chronicles:

> At the age of 21 I was offered union with a woman. I achieved this with the greatest possible difficulty, experiencing the most acute pains of the bladder.

This somewhat less than magic moment seems to be the extent of his youthful amorous adventures. Solace from a miserable life was not to be found in love.

His appraisal of his family and fellow students is harsh, but his most unsparing assessment is of someone else:

Johannes Kepler, the lapdog from Weil der Stadt.

> That man has in every way a dog-like nature. His appearance is that of a little lapdog. Even his appetites were like a dog; he liked gnawing on bones and dry crusts of bread, and was so greedy that whatever he saw he grabbed; yet like a dog he drinks little and is content with the simplest food.... He is bored with conversations, but happily greets visitors like a dog; but when something is snatched from him, he sits up and growls. He hates many people exceedingly, and they avoid him.... He barks at wrong-doers. He is malicious and bites people with sarcasms.... He has a dog-like horror of baths....

This is how the twenty-six-year-old Kepler described himself in the third person. Whatever psychiatrists might read from this, a physical scientist recognizes a born observer of nature.

✦ ✦ ✦

As part of the broad curriculum required by the German universities, Kepler was taught astronomy at the University of Tübingen by one of the leading astronomers of the day, Michael Mästlin, from whom he would have learned of the models of Ptolemy and Copernicus. Kepler was always a Copernican. Just as Tycho Brahe never doubted that Earth stood still, Kepler never doubted it moved.[2]

Whereas the noble Danish astronomer king Tycho Brahe seemed driven to astronomy as if by heavenly intervention, Kepler's route to astronomy seemed to have been a complete accident. While studying theology at the University of Tübingen, Kepler was offered a position as teacher of astronomy and mathematics in Austria at a school in Gratz. This must have been quite a surprise, because he wasn't particularly interested in either mathematics or astronomy. His interest was in becoming a Lutheran minister. It seemed that although his genius was recognized by his teachers, they were anxious to be rid of the annoying, barking lapdog, and recommended him for the position. Had he not inherited a bit of an adventurous spirit from his father, perhaps he would have declined the offer. Whatever his reasons, he accepted the offer with the stipulation that he could return to theological studies if

[2] Kepler's instinctive belief in the Copernican system was not a result of Mästlin's prejudice. An indication of just how far Kepler was ahead of his contemporaries is that even after Kepler's demonstration of the laws of planetary motions, his old professor Mästlin wrote a book about Ptolemaic astronomy.

things didn't work out. But work out they did, beyond the imagining of anyone, Kepler included.

THE COSMIC MAGICAL MYSTERY TOUR

Kepler the teacher was every bit as popular as Kepler the student and as successful as Kepler the lover. He stumbled and mumbled through his first year lecturing before a handful of students. He entered the classroom for his second year of lectures only to find it completely empty; not a single student had signed up for his course. Could the students have any inkling that the bumbling lapdog of a professor was one of the greatest intellects of the century? Luckily, in the happy days of the end of the sixteenth century, deans and college administrators were more enlightened than many of their modern counterparts, and they kept Kepler despite his shaky start as a teacher. They allowed him to teach a course on Virgil in addition to his sparsely attended course on mathematics, noting that "the study of mathematics is not every man's affair," a statement echoed, although not as eloquently, by many modern college students.

Clumsily at first, Kepler mastered astronomy and mathematics, and managed to attract a few students. Then something miraculous occurred. On July 9, 1595, while plodding through a lecture on mathematics, Kepler was stunned by what he felt was the greatest thought of his life, a solution to a question that had been troubling him for some time. It is impossible for anyone to describe his feeling at that moment. He himself wrote, "The delight I took in my discovery I shall never be able to describe in words." In one single moment, Kepler thought he had an answer that would lead him to revolutionize astronomy as no one had before or has since.

To arrive at a great answer, one must come up with a great question. The profound question Kepler was ruminating on that momentous day in 1595 was "Why are there six planets?"[3] Why not just one planet? Why not 137 planets?

That Kepler asked the question at all is a clear indication of his genius. Here was a person who was unwilling to accept the idea that the universe is just "as it is," but wanted to know *why* it is as it is. He certainly had no interest in becoming an astronomer merely to "save the

[3] In the Copernican system of 1595 there were six planets: Mercury, Venus, Earth, Mars, Jupiter, and Saturn (the only planets visible to the unaided eye).

appearances" by building toy mathematical models to describe the motions in the heavens. He wanted to know *why.* The truly amazing thing is not that Kepler arrived at an answer to the question, but rather that he arrived at the question.

Every scientist can recognize the process that led Kepler to his solution of the "why six" question. Newton wrote that he solved problems "by thinking on them continually," but mere mortal scientists solve problems in a different way. We think about them for a while, get nowhere, become frustrated, and put the problem in storage in the back of our minds. Occasionally we will retrieve it from storage, chew on it a bit, ask it in a different way, look at it from another angle, perhaps make progress on it, perhaps not, and put it back in storage. The process continues: retrieve it the next day, look at it again, kick it, curse it, turn it upside down, and put it back in memory. The usual way the problem is solved[4] is that when thinking about some different problem, we will see some connection to the original problem we didn't see before. Or perhaps while reading someone else's work in a related area, we will unconsciously see the problem from a different angle and the "obvious" solution will appear.

Kepler had been struggling with the "why six" question for some time and getting nowhere. His solution came to him not during a lecture on the number of planets but in the course of a class on mathematics. The topic for the day's lecture was an aspect of geometry seemingly unrelated to astronomy. Kepler was demonstrating how a triangle can be circumscribed *around* a circle (the smaller circle just touching the center of each side of the triangle) and can be inscribed *into* a circle (all the vertices of the triangle just touching the larger circle). Kepler drew these figures on the chalkboard. Then the chalk fell from his hand as he stared dumbfounded at the figures he had drawn. The students took no notice of the professor staring agape at the board. It was just another vintage stumbling, bumbling Kepler performance that other students complained about in the course evaluations.

What struck Kepler was that the ratio of the diameters of the two circles was (almost) exactly the ratio of the distance from the Sun to the planets Jupiter and Saturn. An elementary exercise in geometry shows that the radius of the outer circle is twice the radius of the inner circle.

[4] In fact the problem is usually not solved. It is often said that the hardest thing in sport is to hit a baseball, since even the best batters fail about 70 percent of the time. Any physicist would be happy to fail only 99 percent of the time.

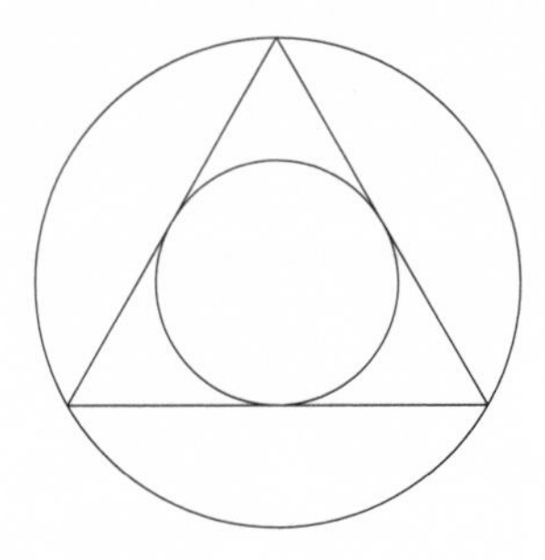

A triangle circumscribed around a circle and inscribed into a circle.

The modern value for the radius of Saturn's orbit is 1.8 times that of Jupiter's orbit, very close to the accepted value in Kepler's time. The number 1.8 must have been in storage somewhere in the back of Kepler's mind, and something clicked when he noticed that 2 is equal to 1.8 (well, at least it's close). The lecture now took a bizarre twist. What began out as a lecture on geometry suddenly became a lecture on astronomy! As Kepler wrote:

> ... the triangle is the first figure in geometry. Immediately I tried to inscribe into the next interval between Jupiter and Mars a square, between Mars and Earth a pentagon, between Earth and Venus a hexagon.... And now I pressed forward again. Why look for two-dimensional shapes to fit orbits in space? One has to find three-dimensional forms—and, behold dear reader, now you have my discovery in your hands!

The "discovery" was based on the fact that in three dimensions there are only five "Platonic" solids that can be both circumscribed around a sphere and inscribed into a sphere. Kepler had the answer! Six planets—five spaces between the six planets—each filled with one of the Platonic solids. That's it! Eureka! Why there are six planets is obvious! Geometry rules! It's all so simple! On the basis of the model one can now even predict the distances of the various planets from the Sun.

The lecture was over (much to the relief of the students). They filed out of the classroom shaking their heads, unaware that Kepler had just embarked on a journey that would lead him to discover what is now known as **Kepler's laws** of planetary motion. Little did they realize they

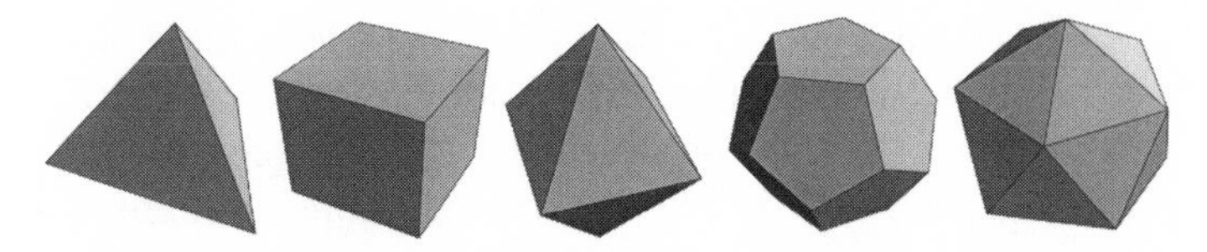

The five Platonic solids: tetrahedron, cube, octahedron, dodecahedron, and icosahedron. All of them can be circumscribed around a sphere and inscribed into a sphere.

were witness to an event that would change astronomy and physics, indeed our world, forever. Perhaps a student stopped and questioned Kepler why a lecture on geometry had ended as a lecture on astronomy. Perhaps someone asked the inevitable question: "Professor Kepler, are we responsible for this on the final exam?"

Within six months of his discovery, the twenty-five-year-old Kepler had written his first book, *The Cosmic Mystery*, presenting his idea to the world. It was written with all the intensity and enthusiasm one might expect from a twenty-five-year-old scientist who believed that a great secret had been uncovered, as if by magic.

There are only three minor problems with the cosmology of Kepler's *The Cosmic Mystery.* The first problem, the only one known to Kepler, was that the ratios of the diameters of the planetary orbits didn't quite work out to be exactly the same as the geometric ratios of the Platonic solids

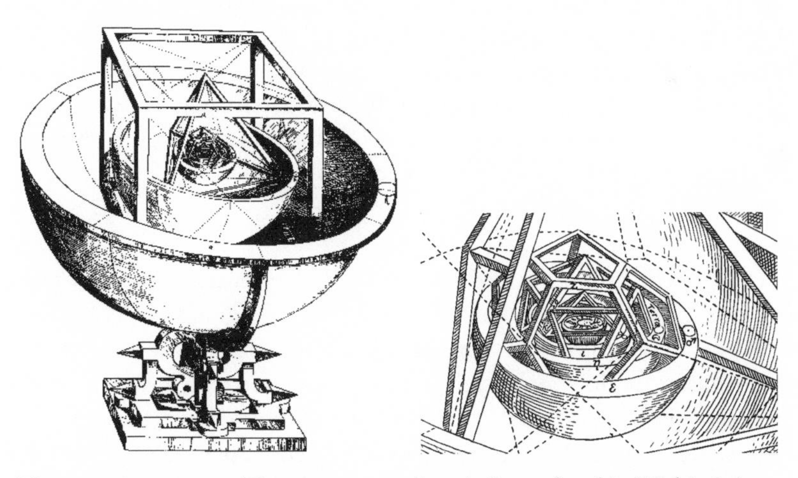

The cosmic mystery. *The outermost sphere is Saturn's orbit. Within it is a cube, followed by the sphere for Jupiter's orbit, a tetrahedron, Mars's orbit, a dodecahedron, Earth's orbit, an icosahedron, Venus's orbit, an octahedron, and Mercury's orbit. The insert details the inner structure.*

(1.8 is not equal to 2). Well, so be it! Rarely can a revolutionary discovery in science neatly accommodate every extant experimental result. Some experiments are incorrect; it is possible that one of the accepted values for orbits, sizes, or whatever, might be wrong. Kepler realized that the model likely needed a little fine-tuning. A couple of screws in his model of the heavens might be loose. Perhaps the discrepancy could be explained if the walls of the solids had an appreciable thickness. Kepler didn't quite know what the problem was, but he was not unduly concerned. After all, he reasoned, how could such a beautiful geometric model come so close and not be correct? The second problem with the model is that we know there are more than six planets. But the planets Uranus, Neptune, and Pluto would not be discovered until the eighteenth, nineteenth, and twentieth centuries.

The third problem with the model is that it is complete, total, and absolute nonsense. Not only doesn't it work out numerically and there are more than six planets, but the whole idea of a planet moving on a sphere inscribed into a dodecahedron, circumscribed around an icosahedron, is simply wrong. To a modern eye, it is comically wrong. Not only that, but from a twentieth century viewpoint the very question Kepler thought to be a question for the ages, why there are six (or nine, or ten) planets, is not an interesting physical question at all.

The Cosmic Mystery was not just the whim of a rash twenty-five-year-old. Twenty-five years later, after he had discovered the three laws of planetary motions, he published a second edition of *The Cosmic Mystery*, further embellishing the idea. Kepler went to his death believing that it was his greatest scientific contribution. Although it was crazy, it guided all his work in astronomy. *The Cosmic Mystery* Tour led Kepler to discoveries to which he seem pulled, as if by magic.

After the publication of his first book, Kepler felt secure enough to look for a wife. But who would marry him? There are many cruel stereotypes: football players are dumb, opera singers are corpulent, accountants are boring, lawyers are unethical, and movie stars are vacuous. Another standardized mental picture is that scientists are nerds. Of course, stereotypes sometimes are on target. Kepler *was* a nerd. Had they been available at the end of the sixteenth century he would have had a plastic pocket protector for his pens, and a piece of white tape securing his thick glasses together. Unlike most modern astronomers, he was not attractive, he did not cut the dashing figure of a Tycho Brahe, and he was miserably poor. Not only that, but women are not generally attracted to someone with a worm on the middle finger of his right hand. Clearly he was not on the "most eligible bachelor" list in Gratz.

A possible candidate for a wife was finally found. Barbara Mühleck was a twenty-three-year-old (clearly well past her prime), twice-widowed daughter of a mill owner. Her parents were eager to see her thrice-married and out of the house for good. In his usual unvarnished style, our inveterate chronicler describes his blushing bride as "simple of mind, and fat of body, with a stupid, sulking, lonely, melancholy complexion." It was obviously a match made in heaven.

But at the last moment a small problem arose. Barbara's father was a proud man. "Sure she's old, ugly, fat, and a burden to the family, but by God, *no daughter of mine is going to marry an astronomer!*" In the end, however, having an astronomer son-in-law did not seem as bad as having an old-maid daughter in the house, and Barbara's father consented to the marriage.

Any happiness and stability Kepler would find was short-lived. There was trouble brewing across Germany and Austria. Kepler was a Protestant, teaching in a Protestant school in a predominantly Catholic Austria. In 1598 the school was closed by the authorities, and under pain of death the professors were expelled from the country. Only one was allowed to remain: Johannes Kepler. It seems that the genius of Kepler the astronomer was recognized by a powerful group within the Catholic Church interested in protecting the great scientist. It was a group that over the next century would weave in and out of the development of astronomy and cosmology, for good as well as bad: the Society of Jesus, or the Jesuits.

The Jesuits were the elite intelligentsia of the Catholic Church. Following the Jesuit motto "Ad majorem dei gloriam" (To the greater glory of God), they saw the greater glory of God in the new developments of astronomy. They were led by the daily recitation of the Spiritual Exercises of Saint Ignatius Loyola, which in part reads "Our one desire and choice should be what is more conducive to the end for which we are created." Perhaps they thought that an understanding of the universe is one of the ends for which we are created. For whatever reason, they protected Kepler. But the protection did not last forever. The Hapsburgs would not spare a single astronomer in their goal to purge Austria of the Protestant heresies. Facing imminent arrest, on January 1, 1600, Kepler departed Austria to become an assistant to Tycho Brahe, the newly appointed imperial mathematician to Rudolph II of Bohemia.

THE DOGS OF WAR

Although the births of Tycho and Kepler were separated by only twenty-five years and five hundred miles, insignificant times and distances on a cosmic scale, it was as if they were of different worlds. Tycho came from one of the wealthiest and most powerful families in Denmark, whereas Kepler came from a decrepit lower-class family in southern Germany. Considering the social structure at the beginning of the seventeenth century, it is almost impossible to imagine the two men overcoming the enormous gulf of class distinction and forming a true friendship. Kepler wrote of Tycho to his old teacher Mästlin:

> Any instrument of his [Tycho's] cost more than my whole family's fortune put together... he is incredibly rich, but he doesn't know how to make proper use of it, as is the case with most rich people.

Negotiating the details of Kepler's appointment was too lowly a task for Tycho, so he dealt with Kepler mostly through intermediaries. What Tycho did see of Kepler, he didn't care for. In a letter to an assistant, Tycho described Kepler not as a lapdog but as a "mad dog." It was truly a collision of people from different worlds, each turning to the other out of desperation.

Relations between the two men started off on the wrong foot, and then got worse. On several occasions Kepler left Prague in a huff, only to write Tycho and beg to be taken back. Other times it was Tycho who dismissed Kepler, only to ask him to return. As is so often the case when people repeatedly quarrel and separate only to reconcile, the attracting force was not respect, admiration, or love, but something stronger: need. Kepler needed the treasure of Tycho's observations to fit together the loose pieces of *The Cosmic Mystery.* Tycho knew that he was an old player in a young person's sport. He needed the ingenuity of Kepler's mind to fit together the missing pieces of his Tychonic model of the universe.

The only common ground on which both Kepler and Tycho stood was that they were the most perceptive watchers of the sky of their time. But the fascinating interplay of the feisty lapdog and the venerable Great Dane in a war of worlds would not last long. They were together only eighteen months before Tycho passed away. But before he died, Tycho set Kepler to a task that would pay spectacular dividends.

One of the most important duties of the head of a research group is to suggest problems and directions for young researchers. The most promising and difficult projects are given to the best assistants. Tycho must have had great hopes (or perhaps great dislike) for Kepler, because he assigned to him the thorniest problem in celestial mechanics: the study of the orbit of Mars. In youthful enthusiasm Kepler accepted the challenge, boasting that he would declare war on Mars and conquer it in eight days. Although Kepler would eventually win, it would take nearly eight years, rather than eight days, to conclude this war of the worlds.

Kepler must have known that Mars had been the match of many astronomers. One casualty of a previous war on Mars was Georg Joachim Rheticus (1514–1576), an Austrian mathematician and astronomer who served as assistant to Nicolas Copernicus in the last four years of the Polish astronomer's life. He is a true hero in the story of the publication of *Die Revolutionibus* (*The Revolutions*), in which Copernicus proposed the heliocentric cosmology that bears his name. It is unlikely that *The Revolutions* would ever have seen its way into print without Rheticus coaxing (*prying* is a better description) the manuscript from the hands of the reluctant Copernicus. While preparing the book for publication, Rheticus spent much time trying to clean up Copernicus's complicated description of the orbit of Mars involving no fewer than five epicycles. As with every astronomer before Kepler, Rheticus found that the motions of Mars defied explanation. Legend has it that he was so frustrated that he conjured up spirits to help him. A spirit dutifully appeared and seized Rheticus, lifted him up, smashed his head on the ceiling, then flung him to the floor, and departed with the admonition "Thus are the motions of Mars." Astronomers of Kepler's time did not believe the story, but many were driven at one time or another to bang their heads on walls in frustration trying to understand the motions of Mars. History is kinder to Rheticus than the spirit was, but Copernicus was not: he failed to mention him or thank him anywhere in *The Revolutions*. Although Rheticus survived Copernicus by nearly thirty years, he never rose to true prominence, and his career was marred by several scandals involving a proclivity for what one German referred to as "the Italian perversion."

Kepler did not resort to summoning spirits to assist him; his only ally was the spirit of his determination. In the end, Kepler emerged victorious from his war on Mars, but the war was not easily won. Before Kepler could attack the problem, he had to escape from the mental confines of three assumptions made by astronomers since the time of Aristotle:

1. All motions of the planets can be described by a series of circles.
2. The planets travel with constant velocity about some point in space.
3. The orbits of the planets are all in the same plane, and that plane passes through Earth (in the Ptolemaic model) or through a point outside the Sun (in the Copernican model).

None of these assumptions is correct. The first two imply that the planets move with *uniform circular motion*, but as Kepler would discover, in fact the motion is neither uniform nor circular. Planets do not orbit the Sun in perfect circles, but rather travel in a path described by an **ellipse** with the Sun at one of the two foci of the ellipse. Planets do not travel

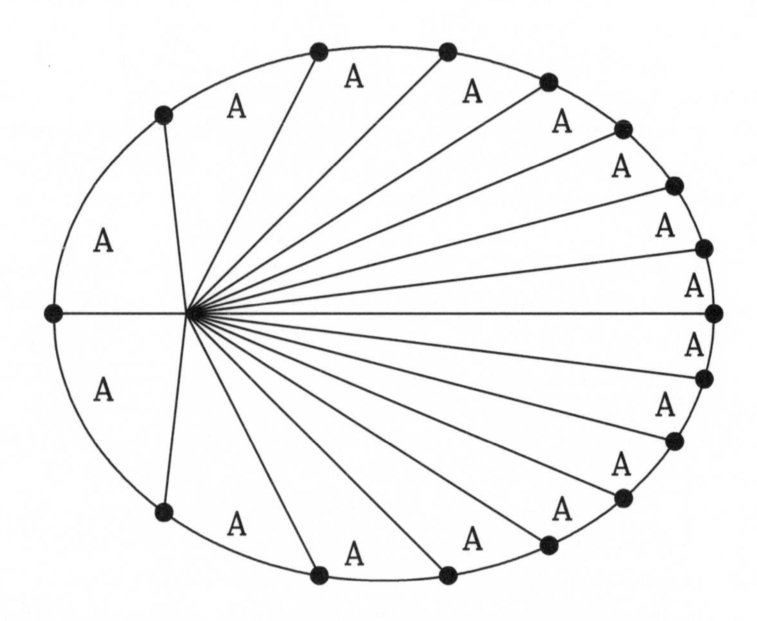

Motion along an ellipse (with a large eccentricity) with velocity increasing as the planet approaches the focus. Equal time intervals along the orbit are indicated by the dots. The velocity increases such that the areas A, formed by connecting the focus of the ellipse and two consecutive points, are all equal.

with uniform speed, but travel faster when closer to the Sun. Finally, the orbits of the planets do not all lie in the *same* plane; each orbit lies in a plane containing the Sun, but the individual planes are inclined by small angles. Although the three assumptions seem plausible, they led astronomers astray for two thousand years.

Copernicus had constructed a heliocentric cosmology employing the "perfect" circle. He wrote that

> it [is] altogether absurd that a heavenly body should not always move with uniform velocity in a perfect circle.

Had Copernicus realized that the price for a heliocentric cosmology was an abandonment of uniform circular motion, he probably would not have proposed the model. No less an iconoclast than Galileo Galilei held on to uniform circular motion and rejected the laws of Kepler. Astronomers of the time would not abandon their sacred assumptions without a struggle. Some of the greatest achievements in science involve discarding unnecessary baggage. It is a courageous step many timid souls fear to take. The first step taken before a revolutionary discovery is often to choose a path never traveled. Kepler the adventurer never shied away from an unknown path.

In hindsight, the assumption of uniform circular motion was due to the imposition of human philosophy on nature. It is simply philosophy to believe that a circle is more perfect than an ellipse because it is more symmetric and pleasing to a human eye. Although symmetry is often considered beautiful and a useful guide in science, nature is the ultimate arbitrator of what is beautiful. As the modern physicist Sidney Coleman said about the development of quantum mechanics, "A thousand philosophers, philosophizing for a thousand years, could not come up with the laws of quantum mechanics." At least a thousand philosophers, philosophizing for two thousand years, had not proposed elliptical orbits for the planets. Nature, as read by patient observation and experiment, is the ultimate philosopher. As a contemporary of Kepler wrote,

> There are more things in heaven and Earth, Horatio,
> Than are dreamt of in your philosophy.

Kepler would never have discovered that the orbits are ellipses if Tycho had not assigned him to work on the planet Mars. Mars has a relatively large eccentricity[5] more than five times greater than that of

[5] The eccentricity is a measure of how much the figure departs from a circle. For more information, see "*The Devil in the Details*" in this book.

Earth's. Next to Mercury and Pluto, it has the largest eccentricity of any planet.[6] It was the "large" eccentricity of Mars's orbit that made it difficult to fit by any combination of circular orbits with epicycles.

Although Mars has a "large" eccentricity, it is important to realize just how "small" large is. If we were to compare the elliptical orbit of Mars with a circle of the same area, we would have to draw the orbit on a fifty-foot billboard just to see a departure of one inch in the separation of the circle from the ellipse. At its closest approach to the Sun (aphelion), Mars is less than 1 percent closer to the Sun than at the farthermost point on the orbit (perihelion).

Mars also has an inclination of its orbital plane of less than two degrees with respect to the orbital plane of Earth. This means that the planet does not always follow the path that the Sun takes across the sky (the plane of the ecliptic). To account for this motion with respect to the ecliptic, all models based on circular orbits had to employ additional epicycles.

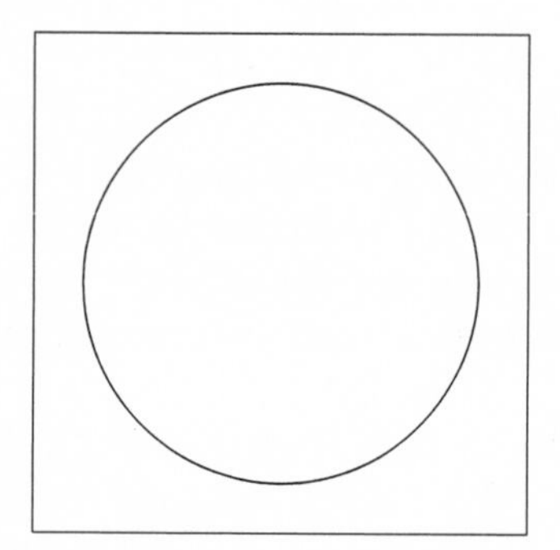

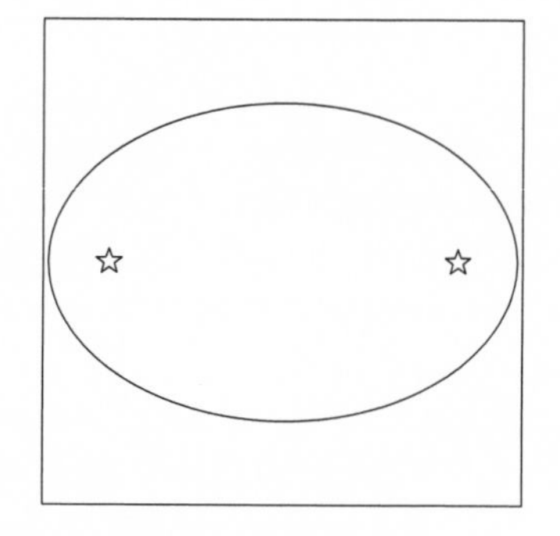

On the left is an ellipse with the eccentricity of Mars's orbit, superimposed on a circle with the same area. There are actually two figures in the left panel, but their difference cannot be seen because the departure of the ellipse from a circle is less than the thickness of the line. On the right is an ellipse with the same area as the circle and ellipse on the left, but with the eccentricity eight times larger *than the eccentricity of Mars. Also shown are the two foci of the ellipse.*

[6] Although the eccentricity of the orbit of Mercury is more than twice that of Mars, Mercury is not often visible because of its proximity to the Sun, so in Kepler's time the observations of Mercury were not as complete as observations of Mars.

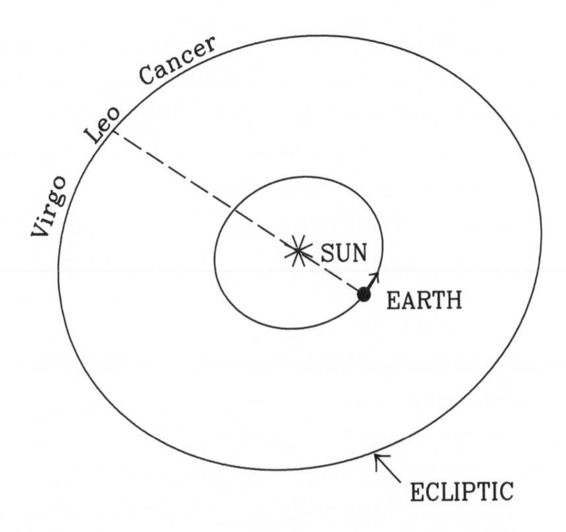

As Earth travels around the Sun, the Sun appears to shift position as viewed against the background stars. The apparent path of the Sun on the celestial sphere is called the plane of the ecliptic. Viewed from Earth, the path of a planet across the sky is near the plane of the ecliptic.

Kepler discovered that no matter how he tried, he had no better luck than Rheticus explaining the orbit of Mars by uniform circular motion. But instead of conjuring spirits or adding a few more epicycles to the model as others had done, he decided to abandon one of the sacred assumptions. And before discarding "circular," Kepler abandoned "uniform." Kepler discovered that if he assumed the orbit of Mars was a circle, but displaced the Sun slightly from the center of the orbit and assumed the orbital velocity of Mars was not constant, he obtained remarkable agreement with observation. After five years of dogged effort, requiring nearly one thousand pages of calculations with the non-uniform circular motion model, he seemed on the verge of the conquest of Mars. What a triumph of human determination and imagination.

Any normal genius would have stopped, published the results, and awaited the accolades from astronomers the world over. But Kepler was not a typical genius. He noticed that his model for the Martian orbit disagreed with two of Tycho's observations by *eight measly minutes of arc.* A circle contains 360 degrees, and each degree can be divided further into sixty minutes, so there are $360 \times 60 = 21{,}600$ minutes of arc in a circle. Eight measly minutes of arc is less than four parts in a thousand of the circumference of a circle.

Until the time of Tycho, no astronomer had even bothered making observations as accurate as eight minutes of arc. It was Tycho's inner passion for accuracy that drove him to make observations of such seemingly superfluous accuracy. He could not have known that the history of astronomy would turn on eight measly minutes of arc. There is no better historical example of the role of accuracy in science. Occasionally precision measurements have spectacular payoffs, even if the potential payoff is unforeseen at the time of the measurement.

The eight measly minutes of arc drove Kepler to abandon the "circular" assumption. Kepler's description of the process that led him to discover that the planetary orbits are ellipses is fascinating. Modern scientific journals report the results of scientific investigations in an impersonal and unemotional manner: state the problem, present the solution, thank the funding agency. Invariably the first paper written by a graduate student is written from a personal and emotional viewpoint, with an elaborate description of all the wrong turns and missteps taken before the final answer was reached ("I wrote a FORTRAN computer program, but it had twelve bugs in it. It took me three weeks to fix them, and my adviser doesn't pay me enough..."). Another duty of an adviser is to return the paper to the student with instructions to convert it into the standard, impersonal style. Perhaps because of his chaotic childhood, the writings of the somewhat neurotic Kepler were anything but stolid.

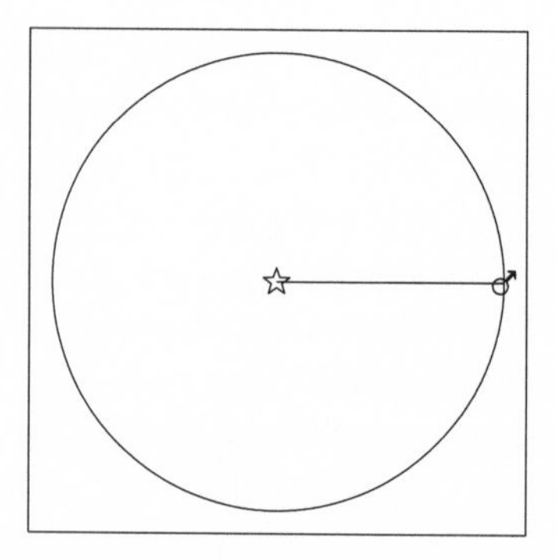

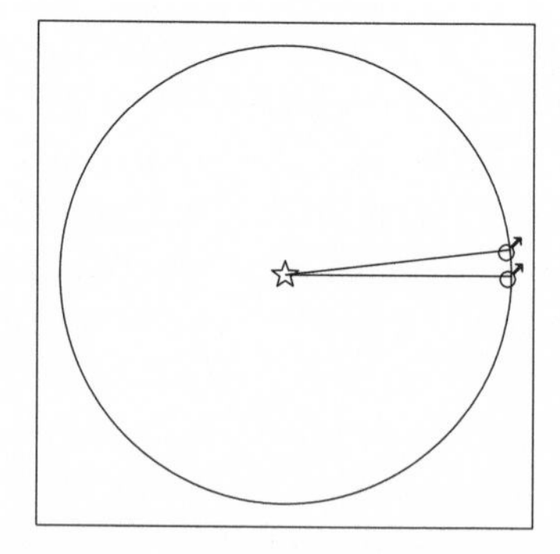

On the left is a figure with two positions on the orbit separated by eight measly minutes of arc. The separation of the two positions cannot be noticed. On the right is the same figure, but with the positions separated by fifty times *eight measly minutes of arc.*

They bristle with personal remarks and bubble with enthusiasm. In *New Astronomy*, his book about the orbit of Mars, Kepler recounts the tour he took in his mind during his 2,400-day wandering in the desert in search of the promised land. He takes us down every wrong path, describes every missed opportunity, and seems to take masochistic joy in pointing out how foolish he was.

Kepler's description of the journey that led him to discover that the orbit of Mars is an ellipse unfolds like a classic Alfred Hitchcock suspense thriller, where we know what the danger is, we see what is waiting around the next turn, but the unsuspecting hero doesn't. We want to scream out a warning (DON'T OPEN THAT DOOR!), but we know the character on the screen can't hear us. So it is for a modern reader as the saga in *New Astronomy* unfolds. We know the answer (IT'S AN ELLIPSE!), but Kepler doesn't.

The saga began when Kepler discovered that his circular model with varying velocity disagreed with two of Tycho's observations by the eight measly minutes of arc. Anyone but Kepler would have ignored this infinitesimal discrepancy; perhaps some would have fudged it away. But just as in the book and movie *War of the Worlds*, where the mighty Martians were defeated by tiny bacteria, in Kepler's war of the worlds Mars was conquered by the minuscule imperfection of eight measly minutes of arc.

The wondrous moment in history when Kepler faced up to the eight measly minutes of arc was the birth of two things. Records state that Johannes Kepler was born on May 16, 1571, in Weil der Stadt. But the records describe only the place and time of his first unassisted breath. The true time of birth of a person is the moment when he or she first looks inward and finds the courage to face something formidable. David's triumph over Goliath has been the subject of many artists. No less a sculptor than Donatello portrayed David's triumphant moment standing astride the dead giant holding the severed head in his hands. But one hundred years after Donatello, a more sensitive, more insightful artist realized that there is a difference between victory and triumph. Felling the giant was the victory, but not the triumph. In the Accademia di Belle Arti in Florence is Michelangelo's portrayal of David's true moment of triumph—the moment when he looked inward and found the courage to pick up a stone and face a giant. When David picked up the stone, he was transformed from a shepherd into a king. When Kepler faced up to the eight measly minutes of arc, he was transformed from a lapdog into a bulldog.

At this point we also have one of the births of modern astronomy, indeed of modern science. At this juncture Kepler realized that science must be exact and quantitative. If a model doesn't fit observations, the discrepancy can't be wished away. Where a model fails and why it fails are important to know. A discrepancy doesn't necessarily mean the model is wrong; it just means the model is incomplete, there is some other effect that hasn't been taken into account in the model, or the data are wrong. No small measure of intuition is required to decide "which of the above." It is a great tribute to the courage and character of Johannes Kepler that he didn't succumb to cheating at this point. After all, after a labor of five years and a thousand pages of calculations, he had arrived at a model that allowed him to find better agreement with observation than any previous astronomer. It is also a great testament to the ability of Tycho Brahe that Kepler refused to believe that Tycho was simply wrong. So Kepler had to face up to the fact that there was something wrong with his model. (IT'S AN ELLIPSE!)

Kepler then took the unimaginable step of turning his back on the wisdom of the ages and rejecting the dogma that the orbit is a circle. But if it is not a circle, what is it? (IT'S AN ELLIPSE!) Kepler soon recognized yet another complication. If the orbit of Mars was not a circle, why should Earth's be? After all, the observations of Mars are made from Earth, so the shape of Earth's orbit must be taken into account. Kepler then asked the imaginative question "What would the orbit of Earth look like as viewed from Mars?"—becoming the first astronomer to leave Earth, at least in his imagination.

First Kepler thought the orbit of Mars was egg-shaped, pinched at one end and fat at the other. But after a year of painstaking work, he concluded that the orbit is some sort of oval. (IT'S AN ELLIPSE!) But what sort of oval? Is there a mathematical equation to describe the "ovalness" of the orbit? Kepler patiently plotted out points in the orbit and tried to determine a geometric equation to describe the oval. He fiddled with the drawing and scratched out calculations over a two-year period with no success. Then he tried something really desperate. Just by patient trial and error he searched for an equation to fit his oval curve. (IT'S AN ELLIPSE!) Most scientists spend their lives trying to compare a model *to* observations. But here, Kepler tried to extract a model *from* observations. The first approach is the task of a craftsman; the second, the work of a genius. Finally, after a couple of additional years' work, Kepler found an equation to describe his oval. The equation he discovered can be identified by a talented high school student of today as the equation of an ellipse. (YES, YES, IT'S AN ELLIPSE!) But Kepler didn't recognize that the formula described an ellipse! He didn't think it described any geometric form at all, and threw away the equation.

Now suddenly the story changes from a Hitchcock thriller to a Marx Brothers farce. After struggling for seven years, after discovering the answer (IT'S AN ELLIPSE!), only to toss it away and start over, Kepler says, "Let me see if the curve is an ellipse!" It is as if God had tired of teasing Kepler and decided to let His hungry lapdog have the bone that had been dangled above his head for so many years. Kepler then recognized the equation he had previously discovered and discarded was indeed the equation for an ellipse. Gold in the dross! As Kepler wrote, "Ah, what a foolish bird I had been."

After eight long, hard years, the war was over. In the end, it was a complete rout. Undefeated for two thousand years, mighty Mars had met its unlikely match. The little lapdog from Weil der Stadt had defeated the great god of war. Once Kepler had the problem in his jaws he wouldn't let go; he smelled blood and he went for the kill. It was a different dog that emerged from the fight.

The spoils of the victory of the war on Mars were the discoveries of the first two of Kepler's three laws of planetary motion:

> The First Law: Planets travel in elliptical orbits, with the Sun in the plane of the ellipse at one of the two foci of the ellipse.
>
> The Second Law: The velocity of the planet increases as it approaches the Sun, and decreases as it recedes from the Sun.

Not only had Kepler destroyed the edifice of uniform circular motion, but he had constructed another edifice in its place. Perhaps some thought the circle was more beautiful than the ellipse (some even suggested turning the ellipse into a circle dressed with many epicycles), but it was the edifice of the ellipse and Kepler's laws on which Sir Isaac Newton would deduce the forces that move the cosmos.

Some combination of genius, determination, ability, and luck, perhaps mixed with a bit of neurosis, had guided Kepler in the war. Arthur Koestler compares Kepler's performance to that of a sleepwalker. While seemingly asleep, he nevertheless avoided all the pitfalls and dangers, and arrived safely at his destination as if guided by a magical unseen hand or some unconscious inner vision.

New Astronomy, containing the first two of Kepler's three laws of planetary motion, was published in 1609 and dedicated to his patron,

Rudolph II. We regard this work as the triumph of Kepler's life, but he regarded it as just a stop on the Cosmic Magical Mystery Tour.

✦ ✦ ✦

Just one year after the publication of *New Astronomy*, the character of astronomy changed forever. One day, in 1610, news arrived in Prague that an Italian mathematician named Galileo Galilei had taken a Dutch optical toy and pointed it at the sky and with it discovered new stars, new moons, and all sorts of wondrous things never before seen by the human eye. Kepler wrote his esteemed Italian colleague and begged him to send an instrument to Prague, because local spectacle makers could not produce instruments of sufficient power. Galileo declined; he preferred to present his telescopes to rich noblemen as trinkets for their amusements. Instead, Galileo sent Kepler a copy of his book describing his discoveries made with the telescope.

Finally, one day the duke of Bavaria, whom Galileo had honored with a gift of a telescope, visited Prague and lent his trinket to Kepler. At long last, the myopic watcher of the sky could see the heavens clearly. Just five weeks later, however, the duke departed and took the telescope with him. We can only imagine the sad spectacle of the greatest watcher of the sky seeing the duke ride away with what to him was a rich man's bauble but to Kepler was a potential tool of discovery.

The early decades of the seventeenth century were tumultuous times in central Europe, as well as in Kepler's life. Kepler's wife died in 1611, and his patron Rudolph II died in 1612 in the midst of a civil war. Astronomy is impossible when the state is threatened by shipwreck, and Kepler was forced to pack up once again and leave Bohemia.

After another period of wandering the continent searching for a job, Kepler was appointed provincial mathematician in the Austrian city of Linz. The intellectual climate of Linz was indeed provincial in comparison with what Kepler had experienced in the capital of Bohemia. The only similarity with his previous position was that his salary was often in arrears, and at times he barely survived.

After arriving in Linz, Kepler set about finding a new wife. This time he had an easier go of it because he was no longer a lapdog but a bulldog. At age forty two (the ideal age for a man), with a successful (although not well paying) career as secure as any in the tumultuous opening years of the Thirty Years' War, this time Kepler made the most-eligible-bachelor list. In his chronicles he described the eleven candidates presented before him with the same dispassionate eyes with which he observed the orbits of the planets. In the end he chose a twenty-four-year-old because of her promise "to be modest, thrifty, diligent, and to love her stepchildren." Kepler's first marriage had not been happy—his wife always disturbed his work and "not much love came my way"—but he seemed to find happiness with his young trophy wife.

Although cut off from the intellectual ferment of Prague, Kepler remained creative. Despite the fact that in the previous six years he had buried a wife and three children and was forced to flee Prague with the sound of war in his ears, he found solace and harmony in his work, titling his next book *Harmony of the Worlds.* The work was closer in spirit to *The Cosmic Mystery* than to *New Astronomy.* After dragging astronomy from the Dark Ages, he slipped back into a model that a modern eye views as comical. In *Harmony,* Kepler returned to polyhedra, but this time introducing a hierarchy of musical harmonic ratios to describe the velocities of the planets.[7] In the course of his investigations of this crazy idea, however, Kepler discovered the third law of planetary motion:

> The Third Law: The period of revolution of a planet increases as the distance from the planet to the sun, in the mathematical proportion of period squared to distance cubed.

As with *The Cosmic Mystery,* the motivation for the study turned out to be wrong, but it led to something spectacular.

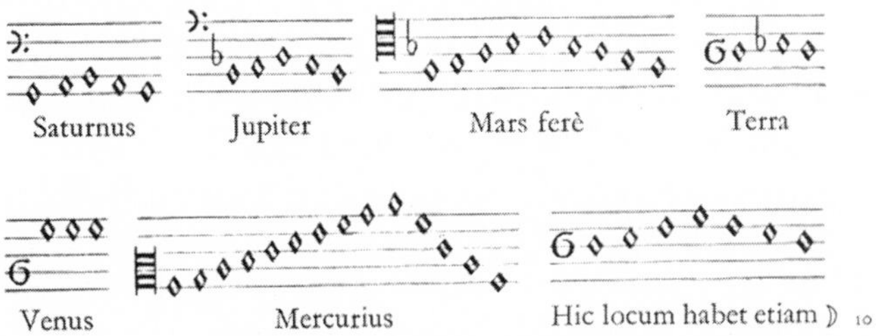

Saturnus Jupiter Mars ferè Terra

Venus Mercurius Hic locum habet etiam ☽ 10

Kepler attempted to unify astronomy, geometry, and music in Harmony of the Worlds *by fitting the angular velocities of the planets to musical harmonic ratios, producing truly heavenly music.*

[7] Modern astronomers are content to leave the music of the planets to Holst.

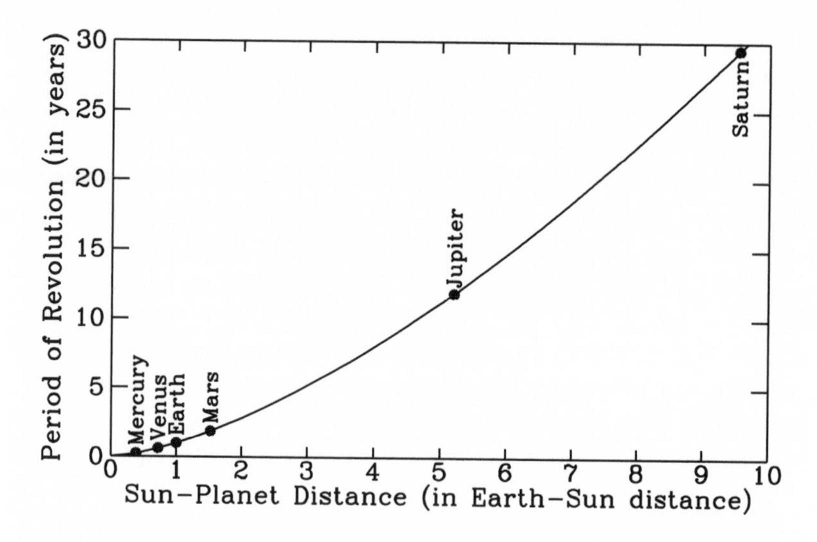

Kepler's third law is a relation between the distance of a planet from the Sun, and the period of its revolution about the Sun. Here is shown the periods and distances for the six planets known to Kepler, along with the theoretical curve obeying Kepler's third law, period squared proportional to distance cubed.

THE GROUND OF ETERNITY

Kepler's relative peace in Linz was not to last for long. In 1626 the peasants revolted and laid siege to Linz, and Kepler was forced to assume the cloak of the wanderer once again. There was no business-class travel in the seventeenth century, and Kepler was a man in his fifties who had to walk instead of ride because of painful boils on his buttocks. He spent time in Ulm and even in Prague once again looking for a position.

While in Prague, Kepler was noticed by Albrecht Wenzel Eusebius von Wallenstein, then at the height of his spectacular military career. Wallenstein was interested in Kepler because he was the greatest astronomer of the time. He could not have cared less about the laws of planetary motions, or *The Cosmic Mystery,* or *The Harmonies of the World.* Wallenstein was a devotee of astrology and wanted Kepler to cast horoscopes for him. He installed Kepler as a private mathematician in the Duchy of Sagan. Neither employer nor employee was satisfied with the arrangement. The problem was that Kepler had too much integrity to pretend he could predict the most auspicious days for battles. So he was relegated to the menial task of providing accurate planetary positions for other more unscrupulous astrologers willing to tell Wallenstein what he wanted to hear.

In August 1630, Wallenstein was dismissed by the emperor, and Kepler lost yet another job.[8] He set out once again to search for a position. He stopped in Leipzig briefly, then went on to Ratisbon. On November 5, 1630, Kepler caught a fever. This time he did not survive.

If Kepler's life were put to music, it would not be an *opera seria* or an *opera buffa*, but a *dramma giocoso* with a sublime mixture of the dramatic and the comical. There was a serious side to his life. On the personal side, there was much personal tragedy and the inner torment of many demons. On the professional side, his work was certainly momentous. The development of the laws of planetary motion is one of the greatest achievements of the human mind. The philosopher Immanuel Kant called Kepler "the most acute thinker ever born." Yet no one, not even Kepler himself, can resist a few laughs at the expense of the little lapdog from Weil der Stadt. Kepler has been the subject of many biographies. Every single biographer feels the irresistible tug of his charm, and in the end loves him and cheers his triumphs.

But as we laugh at the lapdog, we also marvel at how Kepler had grown in his lifetime: from a lapdog to a bulldog! According to some educators, the first step in learning is to establish self-esteem; then accomplishment naturally follows. In Kepler's life, it was his transcendent accomplishments that came first and established his self-esteem. If Kepler had died from his numerous illnesses while in his twenties, he might have left behind the epitaph of a hypochondriac ("See, I told you I was sick"). But by the end of his life Kepler understood the importance of his accomplishments. Knowing that his place in history was secure, he composed this epitaph:

> I measured the skies, now the shadows I measure
> Skybound was the mind, Earthbound the body rests.

In his last hours he did not speak, but continuously pointed his finger, now at his head, now at the sky above him. The Cosmic Magical Mystery Tour was at an end. Johannes Kepler completed the measurement of the skies and commenced the measurement of the shadows on November, 15, 1630. The shadows had never before seen an intellect the likes of Kepler. The shadows were measured with true precision indeed.

[8] It was but a temporary setback for Wallenstein in his roller-coaster career, but it was the end for Kepler.

But Kepler, who had wandered for his entire lifetime, was not to find rest even in death. His bones were scattered to the wind when the cemetery in which he was buried was destroyed in the course of the Thirty Years' War.

In times of uncertainty about the future of physics and astronomy because of political and financial problems, when the advances of our understanding of the universe seemingly meet with hostility from an increasingly large segment of the public, when the prospects for young scientists seem grim, when the future of our field seems beyond our powers to influence, we can learn something from the words of Johannes Kepler in his last letter, written eight days before his death:

> When the storms rage around us, and the state is threatened by shipwreck, we can do nothing more noble than to lower the anchor of our peaceful studies into the ground of eternity.

History often registers the final words of the great. Of little consequence are the final words said *to* those who have shaped our world. In Kepler's case, the last words he heard were from the clergyman Sigismund Christopher Donavarus, who according to a witness "consoled him in a manly way, as behoves a servant of God." We have no idea exactly what Donavarus said to Kepler. But if I may be so presumptuous as to believe Johannes Kepler would be interested in what I might have said, it would be something like this:

> Dear Kepler,
>
> You overcame the most brutal of beginnings. You struggled with inner demons of paranoia and hypochondria. You wandered the continent of Europe barely eking out a living. Despite myopia and multiple vision, you looked into the sky as none before, and became the greatest of astronomers. You discovered the laws of planetary motions that bear your immortal name. You loved your discoveries as you loved your children. But I am afraid that the most beloved child of your imagination is destined to die with you. Your *Cosmic Mystery*, with its baroque constructions of spheres and polyhedra, took you to unexplored parts of the universe. But it was your own private vessel of discovery. Rest in peace in the solace that we will never forget what you have given us. So long as humans

have the curiosity to gaze at the stars, your discoveries will be remembered. But your beloved child will be buried with you. It will not survive as part of our view of the universe. You cherished it, but the light you saw in it was only the reflected light of your other discoveries. You were right about so many other things, but missed the mark on this one. You towered above others of your time and measured the heavens with your marvelous mind, dear Kepler, only in the end to die, as we all do, a blind watcher of the sky.

Rest in Peace

FOUR

The Galileo Equation

Galileo Galilei is the father of modern science. Scientists, historians, and teachers have read, written, heard, and recited that statement so often that most no longer stop to think it might actually mean something. It is a curious thing that people personalize history in such a way. What exactly does it mean to be the "father" of something as vague as "modern science"? If Galileo was the father of modern science, does that imply that his predecessors were not modern scientists? Were his contemporaries "modernized" by Galileo's work, and were those who followed him uniformly modern?

Modern science did not start with Galileo, nor were most people significantly more modern immediately after him. In fact, to think of modern science as having a "birth" may be misleading. Perhaps a better description of science's transition to modernity might be a slow "awakening," for the approach to the study of nature that we call modern science developed slowly in a sequence of fits and starts over a long period.

A millennium or so of the intellectual drowsiness of humanity could not easily be shaken off. It took humanity a long time to clear its collective head of the cobwebs of the past, and even then, periods of wakefulness were interspersed with epochs of stupor. It was as if at various times one person in the crowd would awaken from a deep sleep, rub the sleep from his eyes, look around for a bit, point out something spectacular to see for others who happened to be awake, and then fall back to sleep. Let's take a look at some awakenings in the emergence of modern astronomy. Although others like Regiomontanus, Mästlin, and Longomontanus contributed to the emergence, I will concentrate on Copernicus, Brahe, Kepler, and Newton.

The first of the fab four was Copernicus, who died twenty one years before Galileo's birth. Although he was the father of a revolution that bears his name and the person who "set the earth in motion,"[1] Copernicus's scientific methods were the same as those of Ptolemy in the second century of the Christian Era. In most respects Copernicus was firmly rooted in the Middle Ages. Although he had a revolutionary idea in astronomy,[2] his modernness did not extend to other aspects of his life. He may well have been history's most conservative revolutionary. In addition to studying astronomy and mathematics, he was educated in canon law and as a physician. Surviving records of Copernicus's prescriptions show that he did not practice "modern" medicine, but treated illness with medicines typical of his age: lemon rind, deer's heart, calf's gall, earthworms washed in wine, lizards boiled in olive oil, and that widely prescribed all-purpose magical medicinal elixir of 1510, donkey urine.

In his philosophical outlook, Copernicus subscribed to the philosophy of the Pythagoreans—knowledge is too dangerous to entrust to the rabble of humanity; it should be shared only by a few initiates. In his dedication of *The Revolutions*, Copernicus referred to the letter of Lysis to Hipparchus, which deplored initiating ordinary men into the sacred mysteries of knowledge, and went on to say that "one should not pour the clear water of truth into the muddy wells of the human mind." Copernicus was certainly not interested in leading a revolution where the common man would throw off the confinement of an Earth at rest and view the universe in a whole new perspective. He had had quite

[1] Earth was set in motion at the expense of circles, epicycles, and all manner of complexity—see "The Devil in the Details" in this book.

[2] The heliocentric cosmology of Copernicus was revolutionary, but it was certainly not original. It is possible to trace the idea of a sun-centered solar system at least as far back to the Greek astronomer Aristarchus in the third century B.C. The model was certainly known to the Arabic astronomers of the Marāgana School in northwestern Iran in the thirteenth and fourteenth centuries.

enough of the "commonsense" objections to the mobility of Earth. In *The Revolutions* Copernicus summed up this feeling when he wrote "mathematics is for mathematicians." Although it is easy for a scientist to identify with his frustration at the difficulty of explaining science to the public, it is hard to justify uttering such things, especially in the humanist atmosphere of the Renaissance.

Copernicus awoke from the sleep of the Middle Ages, put the Sun to rest and set Earth in motion, then went back to the comfortable slumber of antiquity.

How about Tycho Brahe? Since Galileo did not seriously consider cosmological issues before about 1608, seven years after Tycho's death, the two men cannot truly be considered contemporaries. Although Tycho was of the generation before Galileo, in many respects Tycho was modern. After all, what could be more modern than to build gigantic, expensive instruments at government expense? Tycho was also certainly modern in his quest to increase the precision of astronomical observations and to replace the haphazard record of data with a complete and systematic survey of the sky. However, in natural philosophy Tycho was an Aristotelian at heart and could not extricate himself from the confines of a stationary Earth in constructing his cosmological model. Tycho never thought for a moment that Earth could orbit the Sun, and the labor of his life, the Tychonic arrangement of the solar system, is in some ways a retreat from Copernicanism.

Johannes Kepler was a true contemporary of Galileo. No one illustrates the awakening of modern science better than Kepler. Kepler's first foray into cosmology was the development of a geometric model for the arrangements of the planets based on spheres nestled within Platonic solids. Although one recognizes a modern yearning to find mathematical patterns in cosmology, there is an unmistakable mystical milieu to the model. After developing this geometric model, Kepler awoke with a start from the slumber of antiquity, discovered the laws of planetary motion in one of the supreme intellectual accomplishments of all time, then returned to the mystical never-land of musical harmonies of the planetary motions in his final major work. It is *not* as if Kepler had one foot firmly planted in modern times and the other foot mired in the past. Rather, he was at some times completely modern and at other times totally ancient. In the parlance of the uncertainty principle of modern quantum mechanics, Kepler was a superposition of a modern state and an ancient state. When he was awake, he saw the sky with modern eyes as none before him, and when he was asleep, his dreams were the mystical ones of the past. Although his genius was manifest in both states, it is difficult to regard him as a modern man.

Another aspect of the lives of Copernicus, Tycho, and Kepler that is strikingly unscientific is their practice of astrology. The fact that the founders of modern astronomy cast horoscopes is something that present-day astronomers would rather keep in the closet. However, their actions should be considered within the context of their times. Although all three did indeed cast horoscopes, in most cases it was expected as part of their job, and they were all too levelheaded to take it completely seriously. Tycho made it clear that he thought that the stars impel rather than compel:

> Astrologers do not bind the will of man to the stars but grant that there is something in man that has been raised above the stars.

As typical in tumultuous times, astrology was very popular during Kepler's lifetime. However, Kepler did seem to step back and forth between the modern scientific disdain of astrology and a mystical belief in it. Kepler called astrology "the step-daughter of astronomy...a dreadful superstition...sortilegous monkey-play...." On the other hand, he did seem to take his own predictions seriously. Although they all practiced astrology to some degree, usually for the money, it should also be appreciated that *everyone* in those times believed in astrology. After all, in the days before genetics, psychology, and brain chemistry, what other explanation was there for the diversity of human behavior? In all periods there are commonly held beliefs, largely unquestioned by those who hold them, that later seem quite absurd. I often speculate which of our cherished beliefs will be ridiculed by future historians.

What about those who followed Galileo? Isaac Newton is considered Galileo's heir in the parade of great scientists. Although the science and the approach to natural philosophy in Newton's two masterpieces, *Optics* and *The Principia*, are the cornerstones of modern science, even the great Newton lapsed into episodes of working on what we would now label as pseudoscience. In the middle and late periods of his life, Newton spent as much time on alchemy, theology, and the study of biblical prophecies as he did on natural science. When awake, he saw the world as no one before him, and with an acuity that possibly will never be surpassed. But his lapses into mystical areas were very much evident.

To understand why Galileo deserves to be called the father of modern science, we must contrast him with the people of his age. While other scientists would awaken from the sleep of the Dark Ages, open their eyes, shake off their drowsiness, and make spectacular advances only to fall back to sleep, Galileo seems to have been born with his eyes wide open—and he hardly ever blinked. He was a thoroughly modern man, with the good, as well as the bad, connotations of the word *modern.*

Although often wrong, sometimes spectacularly wrong, Galileo was always modern.[3] Of all the heroes of astronomy of the sixteenth and seventeenth centuries, Galileo is the only one I can imagine working today in an office down the hall. Transplant him four hundred years to the future, lose the robes, fit him with a pair of (relaxed-fit) jeans, give him a couple of months to catch up on the last four centuries of physics, and he would feel right at home. If anything, it would be difficult for people to adapt to him, for he would always be in your office telling you how dumb you are.

Only in hindsight do we see that Galileo was the father of modern science, for we can appreciate his approach to science in a way that his contemporaries never could. His methods were in many ways too far ahead of his times to be influential during his own lifetime. We now take the scientific methods he developed as obvious and as the only logical way to proceed. It is difficult for us to appreciate how novel they seemed at the time.

A good example that contrasts Galileo's approach to science with the approach of his contemporaries is the dispute over the question of whether projectiles are heated or cooled as they pass through the air.[4] Orazio Grassi, an influential Jesuit scholar, claimed that projectiles become hot as they pass through the air, whereas Galileo said they cool. Grassi was probably more correct than Galileo in this instance, because friction by itself will heat the projectile. But the hapless Grassi used clumsy arguments and made a fatal blunder. (Grassi and Galileo later clashed over the nature of comets; Galileo was mostly wrong about this also.)

[3] It was said of the great twentieth century astrophysicist Arthur S. Eddington that he was "often in error, but never in doubt." While this might have been true of Eddington, it was even more true of Galileo.

[4] The answer is complicated, for there are several competing effects. As a hot projectile passes through the air, heat is dissipated by convection (just as blowing on something hot will cool it), while the friction of the projectile with the molecules of air heats the object. I have also rephrased the issue in modern terms, obscuring the fact that the question is just as complicated as the answer.

To answer this physics question, Grassi did what most people of his age would do: he turned to the authority of the ancients. He quoted a Greek historian who said Babylonians cooked eggs by whirling them through the air in slings. Grassi offered this historical anecdote as physical proof that friction heats projectiles. In the spirit of modern science, Galileo approached the question in a different way. He would not accept blindly the authority of the ancients, but performed the experiment for himself: he took an egg, and whirled it in a sling. What we view as the obvious and simple thing to do did not even occur to Galileo's contemporaries. Galileo whirled the egg until his arm was tired, and then he had his assistants (the equivalent of graduate students) whirl the egg; but for all his trouble he ended up with a raw egg and a sore arm. Not satisfied with that, he boiled an egg and whirled the hot egg. For this situation the convective cooling dominated, and Galileo found a cooler egg after whirling. Galileo reached the obvious conclusion that whirling an egg will not cook it, and in fact will cool it if it was hot to begin with. However, Galileo made the incorrect generalization that friction cools projectiles. Although satisfied in his own mind, he would not be content with simply demonstrating that he was correct; he would be happy only if he could embarrass his foes in the process. In typical vitriolic style, Galileo wrote:

> If he [Grassi] wants me to believe that the Babylonians cooked their eggs by whirling them in slings, I shall do so; but the cause of the effect was very different from what he suggests. To discover the true cause, I reason as follows: If we do not achieve an effect which others formerly achieved, then it must be that in our operations we lacked something that produced their success. And if there is only one single thing we lack, then that alone can be the true cause. Now we do not lack eggs, we do not lack slings, we do not lack strong people to whirl them; yet our eggs do not cook, but merely cool down if they are hot. So since the only thing lacking to us is being Babylonians, then being Babylonians cooks the eggs, and not friction of the air.

That Galileo was not completely correct (perhaps more wrong than right) doesn't really change the fact that the argument between him and Grassi was no contest. Had it been an actual fight, someone in Grassi's corner would have thrown in the towel. Even when wrong, Galileo could humble his opponents and obviously took joy in destroying all those who dared oppose him. His opponents never did figure out what hit them, for they were of the past and Galileo was of the future. It was as if they came to do battle armed with a broadsword, only to find Galileo waiting for them with an assault rifle.

One of the clearest messages in Galileo's work is that a natural scientist should have a healthy skepticism for authority. In Galileo's case it was almost a paranoid distrust of any authority other than his own.

Perhaps Galileo knew in some way that the new experiment-based approach to the study of nature would succeed, but even Galileo might be surprised that it has succeeded so spectacularly. I made this remark to the astrophysicist Dennis Sciama in 1992 during a conference in Venice celebrating the four hundredth anniversary of Galileo's appointment to the University of Padua. Sciama agreed that perhaps no one in the seventeenth century could have predicted that Galileo's methods would work so well, and then went on to ask, "But why didn't someone try them sooner, why did it take so long for the scientific method to develop?" After thinking about it for a couple of years, I am still without a satisfactory answer.

An unstated assumption in Sciama's question is that the natural philosophy we call modern science was an inevitable development. If the present approach to the study of nature is indeed the optimal one, then perhaps there is some Darwinian basis for the assumption. Thus, if one does adopt the viewpoint that modern natural philosophy is optimal, or near optimal,[5] then within the framework of intellectual Darwinism the methods of modern science eventually would have been developed without Galileo. Indeed, we know that by the time of Galileo others were already slowly groping their way toward modern science. But in Galileo's work we find the philosophy of modern science in its nearly final form.

If it is true that there is a final evolutionary end point in the development of natural philosophy—a "right" way to do science, so to speak—then science would be fundamentally different from other human intellectual endeavors. Even if one assumes that given the structure and chemistry of the human brain the development of music and

[5] I don't know of a single scientist who would dispute the statement that modern scientific methods are the best available approach to the understanding of nature. Moreover, it is difficult to find a scientist who believes that one day a better approach to science will be developed that is both more fruitful and fundamentally different from the one employed today. Of course, that doesn't make the original statement correct, since no reasonable scientist confuses consensus with truth.

art is inevitable, it is difficult to argue for the inevitability of any one particular movement in art or music. There is no reason to believe that the development of any single style—say, impressionist art or contrapuntal music—was an inevitable step in the evolution of their respective art forms, nor will there ever be an ultimate expression of art or music in the way that the modern scientific method is the optimal expression of natural philosophy.[6] Of course, even if correct, this doesn't argue that either science or art is better than the other, only that they are different.

In addition to being the father of modern science, Galileo is the most famous martyr of science. The impact of his trouble with the church is difficult to understate; indeed, nearly four hundred years later the effect of Galileo's troubles with the church are still a contributing factor to the uneasy relations between religion and science. To add anything new to the volumes written about the trial of Galileo may be impossible, but it is important to point out that at least part of the blame for the unfortunate confrontation must be laid at the feet of Galileo, for his arrogance played no small part in the tragedy.

Examples of his arrogance are not difficult to find in his writings. For instance, when writing about a competitor, he said:

> You cannot help it, Signor Sarsi [a pseudonym of Grassi], that it was granted to me alone to discover all the new phenomena in the sky and nothing to anybody else. This is the truth which neither malice nor envy can suppress.

It is common today to find scientists who feel that they "own" a particular subfield, and they should be part of any new developments in it. But Galileo felt that he owned the entire universe. As Galileo saw the world, his job was to make *all* the discoveries and the duty of the rest of the world was to praise his discoveries—yet another example of his modernity.

As a theoretical physicist who tries to express the working of nature in mathematical form, I am tempted to express the essence of Galileo's misfortunes in the form of an equation, which I call the Galileo equation:

$$\text{GENIUS} + \text{IMPLACABILITY} - \text{HUMILITY} = \text{TROUBLE}$$

[6] Of course, the product of modern science is different from the tool of modern science; flawed products are often produced by perfect tools.

Galileo Galilei, 1564–1642.

✦ ✦ ✦

The year 1564 saw the death of Michelangelo di Lodovico Buonarroti Simoni, and the births of William Shakespeare and Galileo Galilei. (Indeed, giants walked the earth in those days.)

Galileo was born in Pisa. His father, Vincenzo Galilei, was a composer and a music theorist. As a composer Vincenzo was competent but not exceptional, but as a theorist he played an important role in the development of modern opera. His most notable contribution is contained in *Dialog on Ancient and Modern Music*, a book written in "dialog" form that presents ideas in the course of a conversation between fictional characters. Galileo would later adopt this form for his most influential books. Although one finds many examples of the dialog form in seventeenth century academic treatises, it was quite unusual in the fields of mathematics and science. Perhaps Galileo's adoption of the dialog form is a result of his father's influence.

In light of his later trouble with the church, it is rather ironic that in Galileo's early years he thought of becoming a monk while he attended a Jesuit grammar school near Florence. His father had other ideas, and persuaded him to study medicine at the University of Pisa, where he enrolled in 1581. The university admissions committee did not think much of his promise for future intellectual development, for they admitted him without one of the forty scholarships awarded to children from poor families. The committee's decision seemed justified when Galileo decided that his future was not in medicine and dropped out of college in 1585 without taking a degree.

Galileo then spent a few years at home while his anxious parents watched the young dropout in his early twenties fritter away his days reading or tinkering around building instruments and experimenting with them. But the genius of Galileo survived this rocky start, and his work in experimental physics began to attract attention. In 1589, the twenty-five-year-old Galileo was appointed a lecturer of mathematics at the University of Pisa. His reputation continued to rise, for in 1592 he was appointed to the chair of mathematics at the University of Padua, one of the world's foremost universities. He remained in Padua for eighteen years, the most fruitful period of his life. Although his most important books establishing the foundations for modern physics would not be written for another two decades, most of the ideas were developed during his stay in Padua.

✦ ✦ ✦

While Tycho lived and died believing that Earth was the center of the solar system, and Kepler always seemed to have had an instinctive belief in Copernicus's theory, before about 1610 Galileo seemed noncommittal on the subject. Teaching notes from his days in Pisa show that he was aware of Copernicus's theory but taught Ptolemy's model. In 1597 Galileo wrote both Kepler and the Italian astronomer Jacopo Mazzoni and said that he had become a committed Copernican, but he seemed somewhat less than committed in his writing in the early 1600s. At any rate, he was reluctant to put his beliefs before the public. Possibly his tepid support of the Copernican theory was more of an attack on the physics of Aristotle than on the astronomy of Ptolemy.

Galileo clearly didn't pay much attention to the question. It was not his style to be involved in any issue in which he did not play a central role. His interests in cosmological questions would lie dormant until they were triggered by something that seemed to fall into his lap, as if by destiny.

THE SHARPER IMAGE

Historians may not agree about who invented the telescope, but they are unanimous that it was not Galileo. It was probably invented sometime around 1608 in Holland. All who saw it immediately recognized that the telescope was an important tool, and it wasn't long before its first application was found: the seventeenth-century equivalent of insider trading. With the aid of a telescope, one could get the first glimpse of the ships and cargo coming into port, and obtain a crucial hour or so jump on the trading markets.

Whether Galileo actually examined one of the Dutch "optical tubes" before building one of his own is not clear. Perhaps just hearing that a combination of glass lenses could magnify distant objects was enough for Galileo to put together the idea on his own. One often hears of an idea and immediately says, "Of course, it's so simple, why didn't I think of that." In any case, by the autumn of 1609, Galileo had a crude telescope to demonstrate for the government of Venice.

His telescope caused an immediate sensation with the leaders of Venice, who also recognized that in addition to economic benefits, there would be potential military applications of the instrument. Galileo was quick to take advantage of the situation, and offered the government of Venice exclusive manufacturing rights to the telescope in return for a very large increase in salary. It was characteristic of Galileo's arrogance that he was not particularly embarrassed to sell to the government of Venice the equivalent of the patent rights to something he did not invent and could hardly prevent others from manufacturing.

However much one may criticize Galileo for the attempted expropriation of an invention that was not his own, one cannot escape the fact that possibly, as never before in the history of science, the right tool was in the right hands at the right time, for in 1609 Galileo turned the telescope toward the heavens and changed our view of the universe. Within a few short years Galileo discovered that

1. There are thousands upon thousands of stars that are invisible to the naked eye. This was the first indication that there is more to the universe than we can see, or possibly even more than we can imagine.
2. The surface of the Sun is not perfect, but has small spots. Before this discovery it was assumed that the Sun, like any other celestial body beyond the influence of humans, was unblemished.
3. There are mountains and craters on the Moon.
4. Venus goes through phases just as the Moon goes through phases (waxing, waning, half-moons, quarter-moons, and so on). This would play an important role in Galileo's later argument for the Copernican system.

5. Galileo discovered the rings of Saturn. Although he could not resolve the rings with his primitive telescopes, he could see that the image wasn't round, and described Saturn as a planet with "ears."

6. Galileo noticed that with a telescope he could "resolve" the image of a planet—that is, the planet seemed bigger when viewed through a telescope and he could tell the planet had a finite size. On the other hand, when he looked at stars through the telescope, they still appeared to be pointlike with no discernible size. This suggested to Galileo that the stars must be *much* farther away than the planets. This was also an important piece of evidence he would use in arguing in favor of the Copernican system, for if Earth moves, any nearby object would appear to change position from **parallax.** Since stellar parallax had not been measured, in the Copernican system the stars had to be much farther away than the planets.

7. Galileo discovered that the nebulous area of the sky known as the Milky Way appeared fuzzy because it consisted of thousands of stars too close together to tell them apart.

8. Finally, Galileo discovered four moons of Jupiter, becoming the first person to discover that planets other than Earth had satellites. The discovery that at least one other object besides Earth was the center of an orbital system further strengthened Galileo's belief in the Copernican system.

It is staggering to contemplate how our view of the universe changed within a few short years by the slight improvement of our vision, for it was indeed a relatively small extension of our eyes. Early models of the telescope magnified by about a factor of three, and although Galileo would later claim to build telescopes capable of one-hundred-times magnification, a better estimate for the power of his best instrument might be a factor of thirty. Thus, the very best telescopes that Galileo used were quite crude instruments compared with modern telescopes; in fact, better optics can be found in, say, a $30 pair of binoculars that can be purchased today as a blue-light special in any discount store.

This is the best example of the power of a new tool in confronting the unknown. What lies beyond the horizon of our vision is impossible to predict. In the case of the telescope, there were remarkable discoveries to be made by the extension of the light-gathering power of the eye by a mere factor of three or so. One just can't predict the potential scientific

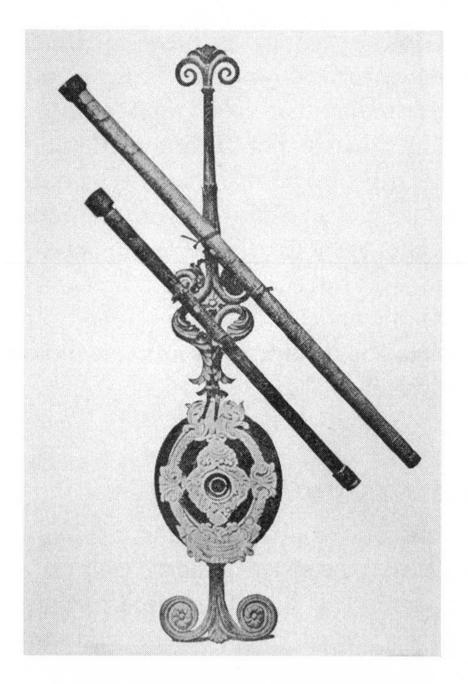

Some of Galileo's early telescopes.

payoff of new scientific instruments, nor is it possible to say how much farther we must search before seeing new phenomena. All we can do is look—look with all our might, with all the resources at our disposal, with all the ingenuity we can muster, and with the hope that it will be fruitful. Most astronomers follow the wisdom of the American philosopher Yogi Berra, who observed, "You can see a lot by looking." Although looking is no guarantee of seeing something, not looking will always ensure not seeing.

Not only did Galileo see what no one else had seen, but he saw that Copernicus must be right. The discovery of the moons of Jupiter orbiting about a body other than Earth seems to have settled in his mind the issue of the Copernican versus Ptolemaic world systems.

Within a year after the publication in 1610 of *The Starry Messenger*, which described his discoveries with the telescope, Galileo was the most famous astronomer in Europe. Kepler's laws might have had a greater impact on the development of astronomy, but Galileo's short (sixty-page) book of his discoveries was easily comprehensible to the educated public, and his discoveries had a wider influence on the development of intellectual thought in the early seventeenth century.

Galileo was quick to parlay his growing fame into a better academic position. He yearned to return to Tuscany, and thought it might be a good idea to name the moons of Jupiter the "Medicean Stars" to ingratiate himself into the favor of Cosimo di Medici, the grand duke of Tuscany. Of course, the modern astronomy community would not allow such an obvious ploy to gain favor with a president or prime minister, at least not without first flattering a few members of the legislature.

Galileo's maneuver was successful, and in 1611 he accepted a position as personal mathematician and philosopher to the grand duke of Tuscany, which included an ideal appointment (no teaching responsibilities) at the University of Pisa.

FOOLISH AND ABSURD

There are many misconceptions about the church's involvement with astronomy at the end of the sixteenth and beginning of the seventeenth centuries. For example, it is widely believed that Copernicus delayed publication of *The Revolutions* until he was near death because he feared reprisals from Rome. But in fact, high officials in the Catholic Church encouraged Copernicus to develop his model for intellectual as well as practical reasons. One practical reason that astronomy was important to the church was its need to predict the date of the first Sunday after the first full moon after the spring equinox—the date of Easter. It should also be noted that Copernicus dedicated his book to Pope Paul III.

To regard the church as a monolithic, narrow-minded entity, with a goal of systematic suppression of knowledge, is also naïve. At the time of Copernicus there were many educated humanists within the church who were truly interested in the development of science and technology. When Copernicus died in 1543 it was still in the early stages of the Reformation, and the church had not yet adopted the fortress mentality that any challenge to her authority must be crushed. Not until 1616, ironically brought about by the actions of Galileo, was Copernicus's book suspended until "corrected" by the church,[7] and not until after the Galileo drama had played out in 1633 was Copernicanism officially declared heretical. In fact, Martin Luther and John Calvin were much

[7] The corrections were quite minor, in keeping with the view of the church that the heliocentric system must be taught as theory, not fact. The decree was not taken very seriously outside of Italy. Of about four hundred copies still in Europe, only about thirty were censored.

more vocal opponents to Copernicanism than anyone in the Catholic Church at the time.

If anything, Copernicus feared the reception that his ideas would have in academic circles. As he put it, he did not want to be "booed and hissed off the stage" by his astronomer colleagues. The academic establishment was very conservative, controlled by devoted Aristotelians who were much more interested in preserving the status quo (and ensuring the continuation of their authority) than in the development of any revolutionary ideas.

The political situation had changed by the time Galileo turned to cosmology in a serious way. It was by then the height of the church's battle against Protestantism, and the church was much more paranoid about new revolutionary ideas. Astronomy was something it was not eager to face up to, for there was somewhat of a minor schism within the church about how to deal with the issue. Although the typical parish priest might disagree, it had become clear to several leading intellectuals within the corridors of power in the church that Copernicanism was the model of choice. We have no way of knowing what went on behind the scenes, but it is reasonable to surmise that the Church planned an "ordered retreat" from the Ptolemaic to the Copernican worldview, perhaps adopting the Tychonic model as a base camp along the way. Given the tenor of the times, it is certainly conceivable that the general feeling within the church was that a radical step in any direction would be dangerous. To those in power, whether it be secular or ecclesiastical power, there are always some truths perceived as too dangerous for public consumption, and they feel it their duty to protect the public from these truths.

At the time of the publication of *The Starry Messenger*, when Galileo's interest turned to cosmology, the church had no official position on the arrangement of the universe. The more cautious within the church hierarchy saw no need to take any stand at all, and were content to allow the teaching of any cosmological model, so long as it was taught as *theory*, not *fact*. The more enlightened among the intellectuals even wanted to avoid at all costs any fundamentalist stand on the interpretation of scripture with respect to cosmology. As Caesar Cardinal Baronis wisely put it, "the purpose of scripture is to teach how one goes to heaven, not how heaven goes." With this background, let's take a look at the situation Galileo faced around 1613.

Galileo was in rather good graces with the authorities in Rome. During a visit there in 1611, he had several friendly audiences with Pope

Paul V, he was elected to membership in the elite Accaèmia dei Lincei, and he was honored and fêted at the Collegio Romano, the Jesuit Roman College that was the intellectual heart of the church. The foremost Jesuit astronomer, Christophe Clavius, professed to be an admirer of Galileo, and said as much to the head of the Collegio Romano, the powerful and controversial Roberto Cardinal (later Saint) Bellarmine, General of the Society of Jesus, Consultor of the Office of the Holy See, Master of Controversial Questions at the Jesuit College, the most respected Catholic theologian of his time, author of *Disputations* (which later was added to the *Index* of forbidden books alongside those of Copernicus and Galileo), one of the Cardinal Inquisitors at the trial of Giordano Bruno,[8] and next to the pope, the most influential person in

Roberto Francesco Romolo Cardinal Saint Bellarmine, 1542–1621

[8] Giordano Bruno was burned at the stake as a heretic in the ironically named Campo di Fiori (Field of Flowers) in Rome on February 16, 1600, for professing ideas about astronomy and religion that were not acceptable to the church. As in the case of Michael Servetus, burned at the stake in 1553 in Geneva by the Calvinists, the real problem was with the religious beliefs, not the scientific views.

Rome. The only point of disagreement about the degree of Bellarmine's power is whether "next to the pope" should be removed from the above description. Galileo also could count on the considerable influence of Cosimo di Medici, duke of Tuscany. This seemed to give Galileo a false sense of security to do and write as he pleased. As part of his arrogance (see the Galileo equation), he underestimated the level of resentment among the many people he had clashed with over the years. Galileo didn't realize that around every dark corner there would be a stiletto wielded by someone once cut to pieces by his barbs.

The church at that time was a model of restraint and tolerance compared with the academic community. Galileo's most implacable enemies were not clerical, but those in the universities who would broach no challenge to the authority of Aristotle on physics and astronomy. If Aristotle said nothing about the moons of Jupiter, if Aristotle said that Earth is the center of all celestial motions, then the Jovian moons were not there and Earth was fixed. Some even refused to look through a telescope to see for themselves. Why bother observing nature when you already know the answer? Those who watch the sky can be excused for occasional blindness (after all, who is not among them?), but those who refuse even to look can only be despised. When one of Galileo's disciples, the Benedictine Father Castelli, was appointed to a chair at the University of Pisa, he was forbidden by the administrators of the university, which was dominated by Aristotelians, to sully the minds of young impressionable students with the subversive notion that the planet Earth moved.

So the situation between the church and astronomers before about 1613 can be described as an uneasy truce. Scientists were free to teach whatever they wished so long as they claimed to be teaching only theories, not facts, and they avoided any scientific interpretation of theology or Scripture. In return, the church would not elevate any astronomical model to the status of official dogma. Tragically, the truce was violated by both sides. Perhaps this uneasy, unwritten truce was untenable.

The first shot across the bow was fired by Galileo. Things started innocently enough at an after-breakfast conversation including Galileo's friend Father Castelli and Grand Duchess Christina, née Maria Madalena of Austria, mother of the duke of Tuscany. Castelli reported to Galileo (who was not present at the dinner) that Duchess Christina argued that Earth could not move because it was expressly declared otherwise in Scripture, particularly in the passage in Joshua 10:12–13:[9]

> ...and he [Joshua] said in the sight of Israel, Sun, stand thou still upon Gibeon; and thou, Moon, in the valley of Aijalon.
>
> And the Sun stood still, and the Moon stopped, until the people had avenged themselves upon their enemies. Is not this written in the Jasher? So the Sun stood still in the midst of heaven, and hastened not to go down for about a whole day.

Although most of the Book of Joshua concerns conquering enemies and avenging defeats (the author of the Book of Joshua seems to have great fondness for the word *smote*), this passage appears to be a clear statement about astronomy; namely, that the Sun goes around Earth, and to extend the hours of daylight it is necessary to stop the Sun from orbiting Earth, *not* to stop Earth from rotating.

However, Galileo had a different interpretation for this passage. He believed (correctly, but without much evidence) that the Sun rotated on its axis just as Earth does, and (incorrectly) that this rotation of the Sun somehow produces a force capable of reaching out and pulling Earth along in its orbit about the Sun. Galileo argued that the biblical passage about Joshua is a description of God stopping the rotation of the Sun, which then brought Earth to rest in its orbit about the Sun, and also caused Earth to cease its rotation. Although Galileo is completely wrong about the physics (the theology also seems a bit confused), it must be appreciated that the concepts of inertia, force, angular momentum, and such things we take for granted would not be developed until Isaac Newton came on the scene later in the century.

Galileo immediately wrote to Castelli in reply, and expanded it into a longer letter to the dim-witted duchess. This letter to Christina was probably the most ill advised action in Galileo's life, for it put in motion the events that would prove not only disastrous for Galileo but even more tragic for the credibility of the church.[10] In the letters to Castelli and Christina, Galileo crossed the line by writing that Copernicanism

[9] Also noteworthy in this context is the passage in Psalms 93:1, which reads, "Indeed the world is firmly established, it will not be moved." Also passages in Psalms 104:5 and Ecclesiastes 1:4–6 seem to support a Ptolemaic cosmology.

[10] Actually many of the arguments in the letters to Castelli and Christina were anticipated by remarks in *Letters on Sunspots*, published by Galileo in 1613.

should be regarded as "fact," and more seriously, he offered his own interpretation of Scripture to support a cosmological model. He also argued that the Bible must not be taken literally, because it was written in a simplified way, "according to the capacity of the common people who are rude and ignorant." Rude and ignorant seemed to be Galileo's opinion of all who disagreed with him. Galileo obviously intended these letters to be circulated, and they were.

His enemies were all too happy to call Galileo's indiscretions to the attention of the authorities. A scientist teaching Copernican astronomy could be overlooked, but a scientist telling church authorities how to interpret Scripture was quite another matter.

In 1615, Bellarmine wrote to Galileo, advising that the Copernican worldview must be regarded as a hypothesis until such time as definitive evidence for the movement of Earth could be offered, and until such time the commonly accepted interpretation of Scripture that Earth was immovable should be taken.

In his writings Bellarmine made a clear distinction between the actual motion of heavenly bodies and the models constructed by astronomers to "save the appearances." Bellarmine seemed convinced that this is how Copernicus himself viewed his model. This view might have been fostered by reading the preface to Copernicus's *The Revolutions*, which reads, in part,

> For it is the duty of an astrologer [*sic*] to compose the history of the celestial motions through careful and skillful observations. Then turning to the causes of these motions or hypotheses about them, he must conceive and devise, since he cannot in any way attain to the true causes.... For these hypotheses need not be true nor even probable; if they provide a calculus consistent with observations, that alone is sufficient.... Let us therefore permit this new hypothesis [Copernicus's model] to become known together with the ancient hypothesis [Ptolemy's model], which are no more probable.... So far as hypotheses are concerned, let no one expect anything certain from astronomy, which cannot furnish it, lest he accept as the truth ideas conceived for another purpose [saving the phenomena], and depart from this study a greater fool than when he entered it.

The preface makes a clear statement that Copernicus's model was simply constructed as a calculational tool and should not be confused with the actual arrangement of the heavens. We now know that the preface is a forgery: it was not written by Copernicus, but inserted by Andreas Osiander at the last moment when Copernicus was too sick to stop it. Unfortunately, this forgery was not widely known at the beginning of the seventeenth century, so perhaps it is not unreasonable that Bellarmine took this view of Copernicus's intentions. This view also appears in the official church decree containing the corrections of *The Revolutions*:

> ... the subject which Copernicus is dealing with is astronomy, whose most distinctive methodology is to use false and imaginary principles for saving appearances....

Galileo went to Rome in 1616 to clear himself of rumors of heresy and to obtain permission to continue to defend Copernicus. Grand Duke Cosimo II instructed his ambassador in Rome to provide for Galileo everything that might be required by a representative of Casa di Medici: rooms at the Villa Medici in Rome for his lodging, a servant to tend to his personal needs, a secretary to handle his correspondence, and a *mulatta* for unspecified purposes. Cosimo also directed his ambassador to account for the expenses, for he expected Galileo to pay for these attendants. It seems that in the complicated genetic makeup of every Medici, whether an enlightened patron of learning and the arts like Lorenzo "il Magnifico" di Medici, or a ruthless despot like the first Cosimo di Medici, the dominant gene was that of a Florentine banker.

The church had been forced by Galileo into taking an official stand on cosmology, for in February 1616, the Father Theologians of the Holy Office, led by Roberto Cardinal Bellarmine, examined two propositions:

1. The Sun is the center of the universe and immovable in local motion.
2. Earth is not the center of the universe and not immovable, but moves as a whole and also with a diurnal motion.

The first proposition was found to be "foolish and absurd, philosophically and formally heretical inasmuch as it expressly contradicts the doctrine of Holy Scripture in many passages, both in their literal meaning and according to the general interpretation of the fathers and Doctors [of the church]." As regards to the second proposition, "All said this proposition receives the same censure in philosophy, and with regard to theological truth is at least erroneous in faith."

The full text of this 1616 decision was not made public until 1633, but the deed was done. The church had taken an official stand on cosmology, although it was not an inextricable position for the church. The decision was that of the Congregation of the Index, and although official doctrine, it was not quite dogma. Luckily for the church, Bellarmine convinced Paul V not to condemn Copernicanism with papal *ex cathedra* authority. Thus the opinion remained that of the Congregation of the Index and did not become officially regarded as infallible dogma, and it remained a position the church could later walk (or back) away from. However embarrassing this action proved to be for the church, it could very well have been a blunder of almost unimaginable proportions had Bellarmine not prevented Paul V from putting the entire prestige of the papacy, indeed the very credibility of the church itself behind the idea that Earth is stationary. If sainthood is a reward for service to the church, then Bellarmine deserved it.

Paul V instructed Bellarmine to inform Galileo that until he had unassailable proof to the contrary, he was admonished to abandon the opinion that Earth moved and the Sun was the center of the universe, that he should neither teach nor discuss the model of Copernicus, and if he did not acquiesce, he was to be imprisoned.

Strangely, Galileo did not seem unduly upset about this turn of events. Ever one to underestimate his opponents and overestimate his ability, Galileo seemed to believe that he would not have any trouble proving that Earth moved.

It is interesting that Bellarmine had an interest in astronomy, and that in his youth Galileo thought of becoming a monk. One cannot help but entertain the delicious irony that had they followed those directions, Galileo might have been the Father Inquisitor and Bellarmine the astronomer before him.

A DIALOG ON THE TIDES

Copernicus had spent a lifetime and failed to develop a proof of Earth's mobility. Galileo surely must have known that this would not be an easy task, for indeed all our senses deny the motion of Earth. If Earth's movement were obvious, feeling it move would not be considered such a special moment. Proof that Earth moved would have to be compelling enough to overcome what sense plainly implied. Galileo realized that the equivalent of a "smoking gun" was needed.

Perhaps at the time of Bellarmine's decision Galileo felt that the tide had turned, that in fact he had the gun in hand, for sometime before 1632 he became convinced that the action of the tides provided the long-sought unassailable proof of the diurnal rotation and mobility of Earth.

Galileo's interest in the tides dated back to his time in Padua. Like many modern tourists, at one point during one of his many visits to Venice while working in nearby Padua, he probably got his feet wet walking across Piazza San Marco in Venice during the high tide. As a person driven by relentless curiosity, Galileo must have wandered to the water's edge to observe the rising and ebbing of the tides. As a natural philosopher, he could not help but ask himself the cause of the movement of such a mass of water.

One characteristic of great scientists is an ability to be enthralled by natural phenomena that others take for granted. Tides had been noticed and studied for millennia, but Galileo's desire was not to predict the size or timing of the tides but to understand the physical cause that moved such a mass of water. Before contemplating the magnitude of the effort necessary to raise the sea a dozen feet, one should carry a few dozen five-gallon water bottles up a flight of stairs. Perhaps then the required effort can be appreciated.

Galileo was not the only person of his age to consider the question of the cause of the tides. Johannes Kepler had also wondered about the tides, and proposed that they were caused by some sort of gravitational force emanating from the Moon. Although Kepler's ideas about gravity were crude, he had essentially the right idea, even if some of the details were wrong.[11]

Galileo dismissed Kepler's theory of a lunar origin for the tides as sheer superstition. Galileo could not believe that the tremendous amount of water being raised and lowered was the result of an invisible force coming from the Moon, and furthermore this mysterious unseen force could pick up millions of tons of water without disturbing so much as a grain of sand on the beach! Clearly Galileo thought that idea was preposterous and that it belonged in the Middle Ages.

Galileo believed that an explanation for the movement of so much water could only involve the movement of an even larger mass, Earth. Galileo's theory of the tides is very simple. Earth rotates on its axis as it revolves around the Sun, so once every twenty-four hours (at midnight) the rotational velocity of Earth will be in the same direction as the

[11] One of the minor details incorrect in Kepler's ideas about gravity is the *sign* of the gravitational force: he imagined it to be a repulsive, rather than an attractive, force.

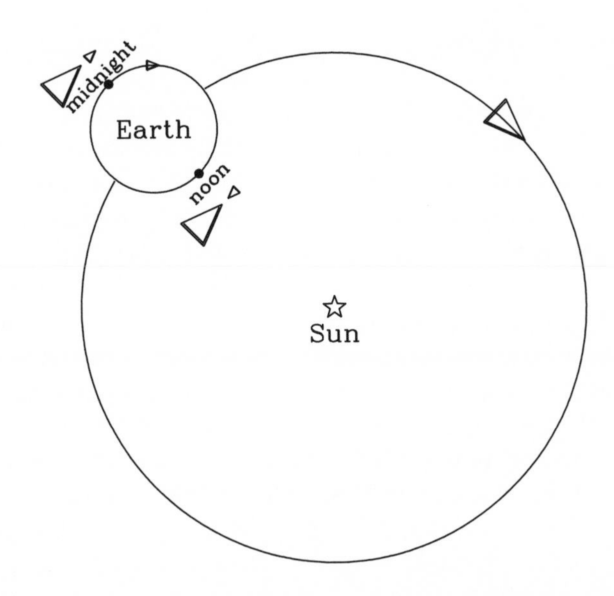

Looking down on Earth's orbit, the rotational velocity of Earth (indicated by the smaller arrow) either adds to or subtracts from the orbital velocity of Earth about the Sun (indicated by the larger arrow). At local midnight the velocities add, and at local noon the velocities subtract. Therefore, with respect to the Sun, *the velocity of any point on Earth's surface is greatest at midnight and smallest at noon.*

orbital velocity of Earth about the Sun, and twelve hours later (at noon), the rotational velocity will be in the opposite direction. Galileo reasoned that this change in velocity causes the ocean to slosh about, much as water in a bowl will slosh about if it is moved back and forth.

Galileo was amazed by the tides, but he was always most amused by his own cleverness. He was so taken by his explanation of the tides that he was willing to bet the universe on it. He thought that his theory of the tides was the smoking gun that could be used to prove beyond a doubt that Earth moved. The explanation is indeed simple; it is also simply wrong.

There are two grounds on which the theory can be criticized. The first is that it is a confused statement about **inertia**. Galileo is often credited as the first person to develop the idea of inertia. In the physics of Aristotle, the "natural" state of motion of an earthly body is rest. Aristotle believed that any body in motion would eventually come to rest. The principle of inertia, first developed by Galileo and later by

Renè Descartes, and perfected still later by Newton, is that a state of motion with uniform velocity is as natural as a state of rest.

Although Galileo might be called the father of inertia, he developed a strange concept known as "circular inertia," which implied that once started, uniform circular motion is a natural state of motion. What we now know as Galilean inertia is the tendency of any body to travel in a straight line unless acted on by an external force. The reason the planets do not travel in a straight line, but orbit about the Sun, is that a force acts on them, the gravitational force. Just as Galileo did not believe in a gravitational origin of the tides, he did not believe that gravity kept the planets in orbit. Rather, he postulated that because of circular inertia the planet would remain in a circular orbit in the absence of a force acting on it. Of course, we know this is not correct. If the switch controlling the gravitational force is somehow turned off, the planets would not remain in orbits about the Sun but would fly off in a direction tangential to the orbit at the location when the switch was thrown. Perhaps Galileo's confusion is an indication of the difficulty in sorting out the concepts of inertia, forces, and gravity that awaited Newton.

A second issue on which Galileo's theory can be attacked is that it predicts that in any twenty-four-hour period there will be one high tide at noon and one low tide at midnight. If Professor Galileo didn't know, then surely any Venetian fisherman or gondolier could have told him that the tides occur at different times of the day and night, and that there can be more than one high or low tide in a day.

Although Galileo can be criticized on physical grounds for the inertia problem, this second issue is not so straightforward, and in my opinion Galileo was not far out of line.

One of Galileo's greatest contributions to science is the idea that in any physical phenomena there are primary effects and secondary effects at play. The best example that illustrates the difference between primary and secondary effects is the important discovery that all objects fall at the same rate.

We all know the apocryphal story. Galileo dropped two objects of unequal weight (most likely graduate students) from the Leaning Tower of Pisa, and noticed that they both hit the ground at the same time. This demonstrated that all objects fall at the same rate, and that the force of gravity does not depend on the mass or composition of an object.

But if anyone really tries the Galilean experiment—say, dropping a bowling ball and a feather from the roof of a building—he or she will be sadly disappointed, for a feather and a bowling ball will not hit the ground at the same time. Of course, the reason is the friction of the air. We would say that the primary effect is that the objects fall at the same rate, and that the resistance of the air is a secondary effect. In some cases the secondary effect overwhelms the primary effect, but nevertheless progress in the investigation of nature requires separating the primary from the secondary effects.

The idea of primary and secondary effects is often attributed to John Locke, but Galileo clearly enunciated the idea over a century before Locke's great work. Galileo reasoned that a high tide at midnight and a low tide at noon would occur only in an idealized situation where there is no friction between the ocean and the sea floor, where the sea bottom is exactly smooth, and where the seashore is perfectly regular. Galileo did know that the tide cycle is not correctly predicted by his model, but assumed that the discrepancy was caused by secondary effects. The identification of primary and secondary effects in any physical phenomenon is certainly not straightforward; it requires intuition, insight, and perhaps a bit of luck. Galileo was spectacularly successful in differentiating between primary and secondary causes for falling objects, but missed the mark on the tides. Although he might be criticized for a sloppy application of his own idea of inertia, he is justified (although wrong) in ignoring the problem of the timing of the tide cycles.

Since he was sure he had the smoking gun that proved the motion of Earth, he set about to convince the world. Between the troubles of 1613–1616 following the letter to Christina, and the publication of *Dialog on the Two Chief World Systems, Copernican and Ptolemaic* in 1632, many of the faces had changed. In 1621, Cardinal Bellarmine, Pope Paul, and Cosimo di Medici died. In Florence, Ferdinand II succeeded Cosimo. Although Galileo remained in the service of the duke of Tuscany and Fredrick held Galileo in esteem, compared with Cosimo, Fredrick was a toady of Rome.[12] Any grief over the death of Paul V or his

[12] But like all Medici, Ferdinand II had the heart of a Florentine banker. In 1633 as the proceedings of the inquisition against Galileo dragged on without an end in sight, Ferdinand instructed the Tuscan ambassador to the Holy See, Francesco Niccolini, to make Galileo pay for his lodging at Villa Medici while in Rome. Unlike the grand duke he served, Niccolini did everything he could to support Galileo, and bravely ignored the pettiness of the twenty-three-year-old Fredrick. He replied that he would not discuss the matter with Galileo, and if necessary he would pay the bills himself.

Maffeo Barberini, 1568–1644; Pope Urban VIII, 1623–1644.

successor, Gregory XV, two years later was tempered by the election in 1623 of Maffeo Cardinal Barberini as Pope Urban VIII. Galileo and Barberini had interacted often in Florence, and Barberini displayed a keen interest in Galileo and his theories. Galileo regarded Barberini as a friend and admirer, and thought he was a truly intelligent man whom Galileo could mold to his purposes. Barberini had even once gone so far as to write (very bad) poetry dedicated to Galileo, reading in part

> Or another marvels at either the heart of the Scorpion
> Or the torch of the Dog Star
> Or the satellites of Jupiter
> Or the ears of Father Saturn
> Discovered by your glass, O learn'd Galileo

Furthermore, Urban appointed a former student of Galileo's, Ciampoli, as his private secretary. As for Bellarmine, perhaps Galileo was glad to see him removed from the scene, for he was indeed a formidable adversary. But Galileo might have been better served had Bellarmine still been around, for Bellarmine's integrity could have been a moderating influence in the debacle of the subsequent trial of Galileo. Certainly the church seemed to lack someone who saw the big picture.

In 1624 Galileo again visited Rome, and his reception was every bit as warm as that of his visit of 1611. He had six audiences with Urban, to whom he had dedicated his most recent book, *Il Saggiatore* (*The Assayer*), which among other things contained many attacks on Grassi. Urban apparently thought highly of the book, for he was fond of having it read to him at the dinner table. What was discussed between Galileo and Urban is impossible to know, but by the time Galileo left Rome he was convinced that he was free to write a book on cosmology. Galileo felt that so long as he couched the book as a discussion and made some half-hearted attempt to present the Ptolemaic argument, he would be safe. It was a spectacular misjudgment.

Dialog on the Two Chief World Systems, Copernican and Ptolemaic was finished sometime around 1630. The book is written in the form of a dialog among three interlocutors—Salviati, Sagredo, and Simplicio. Salviati is Galileo's spokesman: wise, witty, easily disposing of the arguments of his counterpart, the Aristotelian Simplicio. Throughout the *Dialog* Galileo distinguishes between Aristotle and the Peripatetics.[13] In the book Galileo professes great admiration for Aristotle, but shows nothing but disdain for the Peripatetics. Galileo's considered Aristotle a worthy opponent in philosophical arguments, but had no stomach for the Simplicios of the world who would look to the authority of Aristotle as revealed truth. Thus the real struggle of Galileo was not so much against Aristotle but against the *authority* of Aristotle. The third character, Sagredo, enters the dialog as an informed, open-minded person interested in hearing both sides and forming an opinion. The *Dialog* can be seen as a struggle for the hearts and minds of the Sagredos of the world.

In the happy days before litigation became a way of life, it was not necessary for Galileo to have the now common disclaimer that there was no relation between the characters in the book and real people. Everyone knew that Salviati, Sagredo, and Simplicio were based on people Galileo knew. Galileo dedicated *Letters on Sunspots* to his friend and benefactor Filippo Salviati. In fact, the book was written at Salviati's villa, where Galileo went to escape life's turmoil and relax and

[13] Aristotle had a habit of walking about while discoursing on philosophy, and the Peripatetics followed him about. Galileo applied the term in a pejorative manner to describe the followers of Aristotle who would slavishly follow whatever he said rather than reason for themselves.

DIALOGO
DI
GALILEO GALILEI LINCEO
MATEMATICO SOPRAORDINARIO
DELLO STVDIO DI PISA.
E Filoſofo, è Matematico primario del
SERENISSIMO
GR.DVCA DI TOSCANA.
Doue ne i congreſſi di quattro giornate ſi diſcorre
ſopra i due
MASSIMI SISTEMI DEL MONDO
TOLEMAICO, E COPERNICANO;
Proponendo indeterminatamente le ragioni Filoſofiche, e Naturali tanto per l'una, quanto per l'altra parte.

CON PRI  VILEGI.

IN FIORENZA, Per Gio:Batiſta Landini MDCXXXII.
CON LICENZA DE' SVPERIORI.

The frontispiece and title page for the Dialog. *Pictured from left to right are Aristotle, Ptolemy, and Copernicus. The title page reads: Dialog of Galileo Galilei Lincean [member of the Accadèmia dei Lincei]; special mathematician of the University of Pisa; head philosopher and mathematician of the most serene Grand Duke of Tuscany. Regarding a four-day meeting covering a discourse about the two chief systems of the world, Ptolemaic and Copernican; proposing without prejudice one view, then the other, on the basis of philosophy and natural law.*

think. Giovanni Francesco Sagredo was a student of Galileo in Padua, and might have been Galileo's closest personal friend. In the preface, Galileo describes Simplicio as an acquaintance, "a certain Peripatetic philosopher whose greatest obstacle in comprehending the truth seemed to be the reputation he had acquired by his interpretations of Aristotle." Galileo went on to say that he would not mention his Peripatetic friend by name, but refer to him as Simplicio, because his Aristotelian friend greatly admired the sixth century A.D. Greek philosopher Simplicius, who wrote commentaries on Aristotle. Never one to eschew a chance for a sarcastic barb, Galileo could not resist the name. It is supposed that Simplicio was actually a composite drawn from the many Aristotelian academics Galileo had encountered over fifty years of bitter disputes.

The dialog among the Tuscan threesome takes place over four days. The discussions during the first three days show Galileo at his best. The first day is devoted to Salviati's position that there is no clear distinction between the physics describing the terrestrial and celestial regions. Salviati proposes that heavenly objects are made of the same material as earthly ones, and that they are subject to the same laws of motion. Although this seems obvious to us, we should not underestimate the courage required to propose that humans studying the dynamics of earthly objects could divine the laws of motion of bodies in the spheres of the heavens. Indeed, Urban's favorite argument against Copernicanism was that God was not bound by the laws of nature, and no matter what we learned from terrestrial experiments, astronomers could never determine the laws governing the motions of heavenly bodies. Galileo had promised the pope he would include this argument in his discussions. Urban's position was well known, for Urban was fond of making public pronouncements of his knowledge and wisdom. Galileo had quite the opposite view of the question, and said as much in his letter to Christina:

> ...nothing physical which sense-experience sets before our eyes, or which necessary demonstrations prove to us, ought to be called into question (much less condemned) upon the testimony of biblical passages.

Galileo proposed that God follows nature, whereas Urban believed that nature follows God.

The topic of the second day is the rotation of Earth. Simplicio recites the litany of objections to the idea that Earth rotated—objections that Galileo (and Copernicus before him) had heard for years: if Earth rotated, then an object dropped from the top of a building would not fall at the foot of the building; if Earth rotated, clouds would be left behind; and so on. Here again, Galileo is triumphant. A lifetime of studies of motion, acceleration, and inertia had armed Galileo with the insight necessary to give Salviati clear and powerful counter-arguments to dispose of the objections against the rotation and motion of Earth. But at this juncture, although Salviati had managed to refute all objections, he was as yet unable to provide a convincing argument *for* the diurnal rotation.

The discussion of the third day is about the arrangement of the universe and whether Earth orbits the Sun. Here the basic cosmological issue of Ptolemy versus Copernicus is presented. Galileo is not content to have his mouthpiece Salviati espouse the ideas of Copernicus; rather, he has him maneuver Simplicio into constructing the Copernican arrangement by his own hand. Galileo was indeed a born polemic. At the end of the dialog of the third day is a very beautiful passage in which Salviati admits that the movement of Earth is not what our senses tell us, but that in modern science, reason rules over sensory perception. He goes on to say that the popular acceptance of an idea is not an indication of scientific truth. The dialog goes as follows:

> **SAGREDO:** [If the Copernican idea is right,] why has it found so few followers in the course of the centuries; why has it been refuted by Aristotle himself, and why even Copernicus is not having any better luck with it nowadays?
>
> **SALVIATI:** Sagredo, if you had suffered even a few times, as I have so often, from hearing the sort of follies that are designed to make the common people contumacious and unwilling to listen to innovation, then your astonishment at finding so few men holding this opinion would dwindle a good deal. It seems to me that we can have little regard for imbeciles who take it as conclusive proof in confirmation of the Earth's motionlessness that the Earth is too heavy to climb over the sun.... There is no need to bother about such men whose number is legion, or to take notice of their fooleries.... Besides, with all the proofs in the world, what would you expect to accomplish in the minds of people who are too stupid to recognize their own limitations? No Sagredo, my surprise is different from yours. You wonder that there are so few followers of the Pythagorean [Copernican] opinion, whereas I am astonished that there have been any up to this day who have embraced and followed it. Nor can I ever sufficiently admire the outstanding acumen of those who have taken hold of this opinion and accepted it as true; for they have through sheer force of intellect done such violence to their own senses as to prefer what reason told them over what sensible experience showed them to the contrary. For the arguments against the whirling of the earth are very plausible... the experiences which overtly contradict the annual movement [of the Earth] are indeed so great in their apparent force that, I repeat, there is no limit to my astonish-

> ment when I reflect that Aristarchus and Copernicus were able to make reason so conquer sense, that in defiance to the latter, the former became mistress of their belief.

Reason so conquer sense. This passage by Salviati is my favorite of the entire *Dialog.* I have never seen a more eloquent statement in anticipation of twentieth-century physics: reason must conquer sense and guide us in the study of quantum uncertainty, time dilation, confined quarks, curved space-time, the expansion of the universe, black holes, wormholes, superstrings, and other counterintuitive exotica of modern physics.

Galileo had saved the clincher for the fourth day: the motion of the tides. Salviati argues, incorrectly of course, that the action of the tides provides concrete evidence for the movement of Earth about the Sun and the diurnal rotation of Earth.

Poor Simplicio has lost every skirmish. In desperation he falls back on an argument

> that I once heard from a most eminent and learned person, and before which one must fall silent. I know that if asked whether God in his infinite power and wisdom could have conferred upon the watery element its observed reciprocating motion using some other means than moving its containing vessels, both of you would reply that He could have, and He would have known how to do this in many ways which are unthinkable to our minds. From this I forthwith conclude that, this being so, it would be excessive boldness for anyone to limit and restrict the Divine power and wisdom to some particular fancy of his own.

Although indeed Galileo had followed the pope's wishes that his argument be included in the book, it was uttered by the simpleton Simplicio. This was a public humiliation, because it was well known that the pope was identified with this position. Although the argument is treated with mock courtesy by Salviati, it must have stuck in the craw of Urban. One might argue about the exact point where Galileo crossed the line that would guarantee trouble, but by this point he is so far over it, the line can no longer be seen.

By 1630 Galileo had finished the book, titled it *A Dialog on the Flux of the Tides*, dedicated it to the grand duke of Tuscany, and set about securing its publication. Of course, before it could be published, Galileo had to obtain the necessary imprimatur from the church. He brought the book to Rome and submitted it to the Dominican priest Niccolò Riccardi, who was charged with censoring the work. Riccardi was a weak man and really out of his depth dealing with the likes of Galileo. Riccardi did what most weak people do when faced by a decision: he delayed. But eventually Galileo pressured him into approving the book for publication, partly by getting a Florentine censor to approve it. As a hollow gesture of conciliation to make Riccardi feel as if he had actually done something, Galileo agreed to his request to change the title of the book.[14] It is ironic that the censors of the church saved Galileo from the embarrassment of naming the book after the most egregious error in the work. Thus the work was renamed *Dialog on the Two Chief World Systems, Ptolemaic and Copernican*, the name we know it by today, and published in 1632 with the approval of church censors and experts.

The book was an immediate sensation, and soon after its publication not a copy was left on the shelves. Part of the reason for its success is that in addition to being great science, it is great literature. Many of Galileo's writings are classics of didactic literature, with the wicked sarcasm of Swift, humor, wit, and an unusual clarity of style that even today is enjoyable reading. Helping the popular appeal was the fact that the *Dialog* was written in colloquial Italian, not in the traditional Latin of learned scientific works.[15]

It is also a work on a subject of monumental significance: the arrangement of the universe. Galileo was aware of the importance of the topic, for he wrote in the dedication:

> The constitution of the universe, I believe, may be set in first place among all natural things that can be known, for coming before all others in grandeur by reason of its Universal content, it must also stand above them all in nobility as their rule and standard. Therefore if any men might claim extreme distinction in intellect above all mankind, Ptolemy and Copernicus were such men, whose gaze were

[14] This suggestion might have originated from Galileo's ex-student and papal secretary Ciampoli, or perhaps even from the pope himself, who had yet to read the book.

[15] For comparison during a similar period in Spain, of 120 academic books published, 115 were in Latin and only 5 in Spanish.

> thus raised on high and who philosophized about the constitution of the world.

Clearly Galileo was after the hearts and minds of the populace. In the margin of his personal copy of a book by Antonio Rocco attacking the *Dialog* written long after the trial, Galileo wrote of Rocco:

> If I had been writing for pedants I would have spoken like a pedant as you do; but writing for those who are accustomed to reading serious authors, I have spoken as the latter speak.

The *Dialog* is one of the most important books ever written. It is a classic of science,[16] ranks high as a contribution to literature, and is an important chapter in the struggle for free expression.

THE TRIAL OF GALILEO

It has been often noted that one measure of a person's influence is the stature of his enemies. By this measure, Galileo was a giant. His enemies were quick to see the challenge, for the confrontational nature of the book was apparent to all. Indeed, in the margin of his personal copy of the *Dialog*, Galileo wrote out in longhand what he dared not have appear in print:

> Take note, theologians, that in your desire to make matters of faith out of propositions relating to the fixity of the Sun and Earth you run the risk of eventually having to condemn as heretics those who would declare Earth to stand still and the Sun to change position—eventually, I say, at such a time as it might be physically or logically proved that Earth moves and the Sun stands still.

Inside the book was more than just the presentation of an idea; there was a clear challenge to authority, a challenge to which the church felt it had to respond. Perhaps the church saw the book as a test for the ultimate supremacy of the sacred over the profane: whether philosophy would be directed by scientists deducing the laws of nature

[16] The *Dialog* is not Galileo's greatest contribution to science; that distinction belongs to *Two New Sciences*, written after the *Dialog*, in which Galileo laid the foundations for the science of mechanics.

from experiment and observation, or by the father theologians deducing God's laws from their interpretation of Scripture. The church immediately banned the book[17] and summoned Galileo to Rome. Galileo pleaded that he was near seventy years of age and too old and sick to travel. The church responded that he could travel of his own volition, or he would be dragged to Rome in chains. Thus the confrontation that started with the letters to Castelli and Christina reached its inevitable end in the 1633 trial of Galileo before the inquisition.

✦ ✦ ✦

Tourists in Rome routinely admire the façade of the Pantheon, then walk a block or so down its east side to the Church of Santa Maria Sopra Minerva, built in the thirteenth century on the ruins of a Roman temple dedicated to Minerva. Many of the tourists stop and notice Giovanni Lorenzo Bernini's unusual statue of an obelisk on the back of an elephant in front of the church in the Piazza della Minerva. If they enter the church they will be able to admire *The Risen Christ* sculpture of Michelangelo and find, among the many tombs lining the walls, those of Saint Catherine of Siena, one of the patron saints of Italy, and of the Dominican friar Guido di Pietro, sometimes known during his lifetime as Giovanni da Fiesole, but known after his death as Fra Angelico because of the angelic serenity of his religious paintings.

A typical tourist may not know that Santa Maria Sopra Minerva was also home to the Dominican order in Rome. During the inquisition the hounds of God were dedicated to the tireless pursuit of heretics, chasing them down as relentlessly as hounds pursue the fox. Something else not mentioned in many tour books is that looking to the right, on leaving the church, one can see a building that has the appropriate view of the north end of Bernini's elephant heading south. This building, the Dominican convent of Santa Maria Sopra Minerva, was the center of the inquisition in seventeenth-century Rome, and where the hounds finally caught the foxes. It was here that the foxy Galileo was finally cornered by the unrelenting hounds.

Perhaps only in Rome can one find within a fifty-yard radius a Christian church built on the ruins of a pagan temple, the physical remains of one great Renaissance artist, a masterpiece of another, an obelisk on the back of an elephant, the tomb of a saint who represented the gentleness of Christianity, and a building that was the center of the

[17] Of course, this increased its popularity throughout the world.

violence of the inquisition—all conveniently located next to a Holiday Inn. Perhaps on leaving the church some tourists cross the piazza and have a coffee at the bar of the Holiday Inn and wonder if there is anywhere else in the world where such monuments to the genius, greatness, gentleness, savagery, and folly of humanity are so juxtaposed.

✦ ✦ ✦

By 1632 Galileo stood before the inquisition, accused of violating the order given to him in 1616 "not to hold, teach, or defend the Copernican system in any way whatsoever." Despite the fact that the inquisition came nowhere close to providing what we view as fundamental rights of the accused, there was an opportunity to present a defense before being found guilty. So Galileo was given a few days to come up with one. The strategy he came up with was either an indication of his desperation or possibly the disdain with which he held the intelligence of his inquisitors. One might imagine the surprised look on the face of the inquisitors when Galileo appeared before them and said that they misunderstood the book, that it was actually a book *in defense of Ptolemy, and against Copernicus!* He went on to say that he had gone back and looked at the book again, and he now freely admitted that in his zeal to provide a fair discussion, he might have inadvertently let slip from his pen some remarks that someone not insightful enough to realize his intentions (in other words, someone as stupid as you are) might interpret as a defense of Copernicanism.[18] Galileo then went on to tell the inquisitors that they need not worry, for he proposed to clear up this minor misunderstanding by adding a few pages at the end of the next edition of the book that would clarify his true intentions.

Although Galileo might be the father of modern science, clearly he has no claim to be the father of modern legal strategy. I suppose that a legal scholar scouring the annals of legal history might find, with difficulty, a more incredulous defense strategy than that used by Galileo, but I have been unable to uncover one. The strategy he adopted makes the infamous Dan White "Twinkie defense" seem a paragon of reasonableness. Certainly Galileo's plan was spectacularly unsuccessful. He was found guilty.

[18] One must admit that it was easy to be misled that Galileo was defending Copernicus, for he referred to those who did not believe that Earth moves as "imbeciles," "mental pygmies," "dumb idiots," and "hardly deserving to be called human beings."

One of the most tragic episodes in the struggle for human freedom occurred when, early on the morning of June 22, 1633, the seventy-year-old Galileo was dressed in the white robes of penitence, led down the narrow corridors of the Dominican convent of Santa Maria Sopra Minerva, made to kneel before the altar, before the father inquisitors, indeed before the entire world, and forced to recite:

> But whereas—after an injunction had been given to me by this Holy Office, to the effect that I must altogether abandon the false opinion that the Sun is the center of the world and immovable, and that the Earth is not the center of the world, and moves, and that I must not hold, defend, or teach in any way whatsoever, verbally or in writing, the said doctrine, and after it had been notified to me that the said doctrine was contrary to Holy Scripture—I wrote and printed a book in which I discuss this doctrine already condemned, for this cause I have been pronounced by the Holy Office to be suspected of heresy—that is to say, of having held and believed that the Sun is the center of the world and immovable, and that the Earth is not the center, and moves:
>
> Therefore, desiring to remove from the minds of your Eminences, and of all faithful Christians, this strong suspicion reasonably conceived against me, with sincere heart and faith I abjure, curse, and detest the aforesaid errors and heresies.

EPPUR SI MUOVE

Galileo escaped the fate of Giordano Bruno, and the Roman public was deprived of the spectacle of another heretic dragged to Campo di Fiori with his tongue pierced by one iron spike and another driven into his palate to prevent the utterance of blasphemies, then stripped of his clothes and immolated on the stake. But at the abjuration of Galileo, the assembled multitude of cardinals, bishops, and the papal court witnessed what they thought was a symbolic triumph of religion over humanism, for it now appeared that the church had exercised its authority over the most famous scientist in the world. In the very spot where Galileo submitted once stood the Roman temple to Minerva, goddess of wisdom. Perhaps as the cardinals filed out of the church they felt that a new way of science based on interpretation of Scripture would be erected over the ruins of Galileo's seemingly discredited scientific methods, just as the Christian church of Santa Maria was constructed over the debris of the pagan temple of Minerva.

The Roman mass of the seventeenth century began and ended with the symbolic act of kneeling. So it was for the Italian Renaissance. Some scholars place the beginning of the Italian Renaissance in 1361, when according to legend Francesco Petrarch knelt before Giovanni Boccaccio and Leonzio Pilato in gratitude for bringing knowledge to Florence in the form of a translation from classical Greek to Latin of Homer's *Iliad.* Although he was far more famous, Petrarch knelt before the scholars to thank them for the gift. It might also be argued that the Italian Renaissance ended in 1633 when Galileo, a famous Florentine humanist in the tradition of Petrarch, knelt before the inquisitors in Rome and was forced to refute the knowledge he had brought to the world.

According to the Galileo legend, after reciting the required statement denying the movement of Earth, Galileo rose from his knees and defiantly said *eppur si muove*—"nevertheless it moves." Although no one doubts that he believed it, no historian imagines he made such a rash statement. Although he was not physically tortured, as part of the proceedings he was informed of the tortures that would await him if he did not cooperate. Confident that he was correct and would be exonerated by history, he also had the sobering knowledge that he was figuratively (and literally) only a few hundred yards away from the stake. We are forever grateful that Galileo did not choose to follow Giordano Bruno to Campo di Fiori, for in the remaining years of his life Galileo produced works that were of immeasurable importance to the development of science.

In addition to being forced to abjure, curse, and detest the errors and heresies of Copernicus, Galileo was sentenced to live out the remainder of his life under house arrest, and forbidden to discourse with students about cosmology, lest he discuss the mobility of Earth. The *Dialog* was placed on the *Index Librorum Prohibitorum* (*Index of Prohibited Books*), which listed books that contained ideas so dangerous that merely reading about them would imperil a person's immortal soul, where it remained until 1885. Galileo was also required to perform the ritual of reading a daily penance.

It is now widely believed that despite his public pronouncements of his sadness at the troubles of Galileo, and his personal assurances to the ambassador of Tuscany that he would do all in his power to help Galileo, the pope played a pivotal role in the proceedings. Galileo also had made implacable enemies of the Jesuit order. A personal friend of Galileo, the Jesuit Grienberger, wrote to a mutual friend, another Jesuit, and said,

> If Galileo had known how to keep himself in favor with this Order, he would now appear famous before the world.

Galileo does not now, and never did require the support of any religious group to appear famous before the world. Anyone who doubts the inevitable triumph of a correct idea, forcibly stated and passionately defended, need only look to the case of Galileo.

Galileo discovered that the arrangement of the universe is a matter of deep significance, and people would not change their view of the order of the universe without a struggle. In the words of Galileo's fellow Florentine Niccolò Machiavelli in *The Prince*, "It must be considered that nothing is more difficult to carry out, nor more doubtful of success, nor more dangerous to handle, than to initiate a new order of things." Of course, it is possible that Galileo never read *The Prince*, for it too was on the *Index*.

Had Galileo written the book in Latin in a less confrontational manner, perhaps he would have been spared censure. But why should he have to! As Galileo later wrote,

> I do not feel obliged to believe that the same God who has endowed us with sense, reason, and intellect has intended us to forgo their use.

Galileo realized that his sentence would never be commuted. When asked by a friend if he thought the judgment would be rescinded, he replied:

> I do not expect a commutation of the sentence because I have not committed any crime. I could expect forgiveness and pardon if I had done so, because it is such offenses which would give a sovereign the opportunity to show his generosity and forgiveness, while a man who is sentenced innocently must be treated without any pardon to prove that his accuser was right.

However Galileo might be criticized for arrogance, for instigating controversy, or for a proclivity for polemic writing, at times his insights into human nature show flashes of genius.

✦ ✦ ✦

Galileo's part in the development of our view of the universe is pivotal, but in a way different from that of Copernicus, Tycho, Kepler, and Newton. One can trace the development of our picture of the solar system as starting with the idea of Nicolas Copernicus, incorporating the observational program of Tycho Brahe that provided the data that led to the set of three phenomenological laws describing the motion of the planets by Johannes Kepler, and culminating with the explanation of the motions based on the laws of physics by Isaac Newton.

But no one saw the big picture as Galileo did. Although the theory he defended as correct was the theory of Copernicus, not the model of Kepler,[19] in a larger sense, it really didn't matter. Whether the motions are circles or ellipses were details that Galileo did not consider important. Galileo's cosmology, like his astronomy, was qualitative and not quantitative. Although his astronomical discoveries were all of the first rank, and any one of them would make him immortal, he was not the systematic, careful astronomer that Tycho was. But science requires both types of observers, those who have flashes of insight and those who watch the sky in a more systematic manner.

Historians of science who delve deeper into Galileo the man find a very human person, who was at many times vain, obstinate, or just plain wrong. In the life of Galileo they discover that even the greatest of scientists are prone to self-delusion. Fans of Galileo seem to gloss over his human frailties, whereas recent fashion has been to concentrate on them. Critics refer to Galileo as a coward, courtier, or plagiarizer, whereas admirers canonize him as a martyr to the cause of scientific and intellectual freedom.

But neither fan nor critic can change the fact that at the beginning of the seventeenth century Galileo saw that there was a new way to comprehend the universe. Galileo's insistence that "philosopher" be added to the customary title of "mathematician" in his appointment to the

[19] Galileo was not one to spend a lot of time studying what other people did. He never did appreciate the importance of Kepler's discoveries. Sadly, the most creative people are often the least receptive.

Tuscan court was an indication of how clearly he realized the significance of the rudimentary new approach to nature he was developing. Galileo knew that the workings of the universe were not to be discerned by applying pure logic to the words of Scripture or the writings of antiquity. Perhaps he wasn't the first or the only person of his time to use a new approach to "natural philosophy," based on mathematics and physics, as tested by observation and experiment. But no one was as eloquent or forceful as Galileo when he wrote in *Il Saggiatore*:

> Philosophy is written in this grand book, the universe, which stands continually open to our gaze. But the book cannot be understood unless one first learns to comprehend the language and read the letters in which it is composed. It is written in the language of mathematics, and its characters are triangles, circles, and other geometric figures without which it is humanly impossible to understand a single word of it; without these, one wanders about in a dark labyrinth.

These are not the words of a coward or a courtier but of an indomitable spirit who found his way out of the dark labyrinth guided only by the faint lights in the night sky. Galileo watched the sky with imagination and vision, and through a most rare combination of genius and insight knew exactly how to read it.

Even the best fathers are not without shortcomings. If we must accept arrogance, stubbornness, invectiveness, and vanity along with genius, so be it. If modern science must have a father, let it be Galileo!

FIVE

Newton at a Distance

Just about every discussion of the history of modern science recounts the familiar fact that in the very same year that the Italian scientist Galileo Galilei died in the Tuscan city of Florence, the English scientist Isaac Newton was born in Woolsthorpe, Lincolnshire. It is almost irresistible to view this happenstance as a symbolic passing of the mantle from one country to another, and from one great scientist to his successor. But like so many other symbolic coincidences, it is constructed of equal parts fact and fiction, and it is at the same time relevant and insignificant.

The facts are that history records the death of Galileo on January 8, 1642, and the birth of Newton early on the morning of December 25, 1642. But, in the immortal words of Ronald Reagan, "facts are stupid things." The stupid thing about these facts is that Newton was born in the year of Galileo's death only because Italy and England were using different calendars at that time. The Italians had adopted the modern Gregorian calendar in 1582, whereas the English still employed the Julian calendar (and did so until 1752).[1] Because of the incompatibility

[1] The basis of the calendar of Gregory XIII originated with the Neapolitan astronomer Luigi Lilio Ghiraldi, and the idea was developed by Christophe Clavius, the same Jesuit astronomer and mathematician in Rome who confirmed many of Galileo's discoveries.

of their calendars, December 25, 1642, in England corresponded to the January 4, 1643, in Italy. The birth of Newton would have been in 1643, the year following the death of Galileo, had not the English regarded calendar reform as a papal plot and stubbornly clung to the outmoded Julian calendar. Or had the Italians not adopted the Gregorian calendar, Galileo would have died in 1641, the year before Newton's birth.

The relevance of the timing of the death of Galileo and the birth of Newton is that the development of the science of classical **dynamics** (the study of motion, acceleration, inertia, and forces) in some sense began its modern formulation with Galileo and reached its ultimate form in the work of Newton. It might be said that Newton finished what Galileo started.

However, in the development of cosmology, perhaps Johannes Kepler, rather than Galileo, should be considered Newton's true antecedent, for it was Kepler who made use of the observations of Tycho Brahe to construct the phenomenological laws that describe the motions of the planets. This quantification of the motion of the planets in the language of mathematics allowed Newton to demonstrate how the laws follow from a universal law of gravitation. So for the thread of the development of cosmology and the synthesis of astronomy and physics, Kepler was a more significant influence on Newton than Galileo.

There are so many Newton legends that it is often difficult to separate fact from fiction. Of some things we can be sure: in the course of his lifetime Sir Isaac Newton was a student, fellow, and later Lucasian Professor of Trinity College of the University of Cambridge; he was the coinventor of calculus, the author of *Philosophiae Naturalis Principia Mathematica* (*Mathematical Principles of Natural Philosophy*)—the *Principia* for short—certainly one of the most important books ever written on any subject, the inventor of the reflecting telescope, and the author of *Opticks (Optics)*; in the political realm he was elected president of The Royal Society, he was knighted by Queen Anne,[2] he was appointed war-

[2] The degree to which Newton's elevation to knighthood was a recognition of his scientific accomplishments is unclear. Queen Anne knighted Newton in 1705 at the urging of his benefactor Charles Montague, earl of Halifax. At that time Newton was standing for election to the House of Commons from Cambridge, and Halifax, needing all the allies he could muster, thought a little publicity would help Newton in the election. It did not—this time Newton finished a distant last in a field of four.

den and later master of the Mint, and he was elected to Parliament in 1698 and 1701; on the personal side he was one of the most complex and interesting men in the history of science, as well as one of the most original thinkers of all time. He was also an unhappy, neurotic man, who often teetered on the verge of nervous breakdowns, and several times did seem to break under the strain of creative fury.

Not much is known for certain about his early life. We do know that the Newtons of Woolsthorpe were part of the rising middle class of the seventeenth century and benefited from the general prosperity of the times in England. Newton's father, also named Isaac, married Hannah Ayscough in April 1641. The Ayscoughs were a reasonably educated family, unlike the Newtons.[3] Hannah's brother had graduated from Cambridge and was an Anglican minister in nearby town of Colsterworth. A few letters between Newton and his mother survive, so we know she could read and write (although not very well).

The circumstances surrounding Newton's birth and early years were not auspicious. The elder Isaac died three months before the birth of his son. Young Isaac was quite a sickly infant, and his very survival seemed to surprise his family. When Isaac was three, the widow Hannah Newton married the sixty-three-year-old minister Barnabas Smith, eventually having three more children by him before he died at age seventy one. Although they were quite well off and had the means to support him, Mrs. Smith left Isaac in the care of her parents, the Ayscoughs, when she moved a few miles away with her second husband and started a new family.

The death of Newton's father and abandonment by his mother have often been proposed as the principal events that shaped his life and his character. However traumatic these early events might have been, in the end Newton did benefit from growing up with the Ayscoughs in an environment where education had a higher value than if he had been raised as a Newton. Although there is little evidence of affection between Newton and his grandparents, they did see that he received an education.

It is tempting to speculate about what would have become of Isaac Newton had his father survived. Perhaps he was simply destined to accomplish great things. Others have pointed out that had Newton's

[3] Sir Isaac Newton, arguably the most brilliant person ever born, was the first literate Newton; at the time of his birth in 1642, no one on his father's side of his family had ever been able to sign his or her name.

father survived, Isaac might well have been raised an illiterate like the rest of the Newtons. But education alone cannot explain Newton's accomplishments.

Yes, Newton did attend school, but it might be argued that the most important aspect of his education was that it didn't get in his way. One can find nothing in his early schooling, either in the curriculum or any mention of a special teacher, that might have fostered one of the most acute intellects of all time. Perhaps the only thing he encountered outside formal class work that comes closest to providing the catalyst for his development was a book. It was indeed a very special book. But it seems that the book that had the greatest impact on Newton as a child was not a school book, not even a science book. Rather, it was a book he discovered in, of all places, his stepfather's library.

Although he was raised by his grandparents, Isaac occasionally visited his mother and stepfather just a few miles down the road. Reverend Smith was an educated man and had a rather large private library in his home for the young, curious Isaac to browse through. As might be expected in the library of a church rector, most of the books were about theology. But Newton found one book on the shelves that made a lasting impression on him. It might be said that the world changed the day Newton pulled the book from the shelf and opened it, for in the book he found something spectacular, something that excited his imagination. The book that Newton found, which had paramount importance to the development of science, was blank. Empty.

While a young man, Barnabas Smith had purchased a book that consisted of a richly bound collection of quality white paper, apparently with the intention of diligently recording in the empty book important ideas and memorable quotes, as well as other gems of wisdom he would find in the books he read. But like many such projects that people undertake with good intentions, it came to nothing. He made notations on the first few pages before losing enthusiasm for the enterprise, and left behind a great expanse of blank pages of paper. In the seventeenth century paper was precious, and a thrifty person would not throw an empty book away but make use of it. It fell to Newton to make use of the paper.

Oh, how Newton made use of the paper! It was on the expanse of unused pages of Reverend's Smith's book, which Newton called the "waste book," that he first developed calculus and penned his first calculations and musings that led to the science of dynamics and the *Principia.*

It might be said that Newton saw the science of physics as a nearly empty book on which great intellects such as Kepler, Galileo, and Descartes with great effort had been able to write but a few passages, leaving behind a great expanse of blank paper. Newton was blessed (or perhaps cursed) with a relentless drive to fill any piece of paper he came across. In his lifetime he was consumed by subjects ranging from the nature of the Christian Trinity,[4] to alchemy, to natural philosophy, to reorganizing the workings of the Mint. He attacked them all in the same manner. He would quickly assimilate what others had done, as if their contributions were but the opening preface; then he would take off on his own, furiously writing the story into the empty book. By the time he departed, he had written into the blank pages of the book the nearly complete story of classical physics.

Isaac Newton in 1689 at age 46, shortly after the publication of the Principia.

[4] Newton's colleagues in the College of the Holy and Undivided Trinity of the University of Cambridge would have been shocked to discover that he was secretly an Arian who did not believe in the Trinity. In fact, he thought that the worship of Jesus as God was tantamount to idolatry.

Newton seemed to be driven by some unknown force. Many scientists find great joy in the development of new knowledge. But possibly no one was more touched with the strange joy of the pursuit of knowledge than Newton. The pleasure of discovery is something easier to experience than describe. Sometimes the joy of the pursuit is the strange feeling that actually a problem is pursuing you. In previous times when hunting was considered a sport, people who hunted dangerous game reported a sense of exhilaration when the distinction between the hunter and the hunted became blurred. So it is in science. Part of the excitement of science is that occasionally one gets the feeling of being stalked by the problem, rather than stalking the problem. As the Danish philosopher Søren Aabye Kierkegaard (no friend of science) put it:

> Knowledge is an attitude, a passion. Actually it is an illicit attitude. For the compulsion to know is a mania; it produces a character out of balance. It is not at all true that the scientist goes after truth. It goes after him. It is something he suffers from.

In the hunt for the truth, Newton became the hunted.

THE EMPTY BOOK

In July 1661, at age seventeen, Isaac Newton entered the College of the Holy and Undivided Trinity of Cambridge University. During his early years at Cambridge he was a "sizar," which is a genteel word for servant. Although his mother and stepfather were quite well off and could easily have afforded to subsidize completely his education, he was forced to work his way through his early years at Cambridge as a servant to other students.

Newton's formal education at Cambridge had as little impact on him as his primary education. It was a different education from that received by his predecessor, Johannes Kepler. About a half-century before Newton entered Cambridge, Kepler entered the German University of Tübingen. While there, he greatly profited from the discipline and intellectual rigors of a German education. Newton also benefited from the very different academic atmosphere he found at the University of Cambridge, which can most charitably be characterized as "laissez-faire," a less kind description of the curriculum would be "ossified." Not much was required of a Cambridge student at the time.

Whereas the external stimulus of religious strife in Germany demanded that their universities produce clergy with the best possible education, university education in England in the 1660s seemed to be without a clear intellectual purpose. The Reformation and Counter-Reformation raging in Germany at the end of the sixteenth century was the same sort of impetus that Sputnik was to science education in the United States in the 1960s and 1970s. Without a similar stimulus in England, university education seemed to languish and to lack direction, leaving students on their own to find their own level. In other words, it was perfect for Newton.

Some students blossom in an unstructured environment, whereas others require a firm, guiding hand. Kepler flourished (at least academically) in the structured German environment, but Newton thrived in the relaxed, gentlemanly atmosphere found at Cambridge. Certainly the university life Newton encountered at Cambridge was an empty book into which a student could write what he wished. Many students, in the style of Barnabas Smith, wrote little.

But Newton was different. He paid scant attention to the prescribed courses of study, apparently never finishing any of the required readings (and apparently not doing very well in his examinations—luckily, the examinations were not taken very seriously either). But studying on his own, from 1661 to 1663 Newton pursued and mastered the works of the greatest mathematicians and natural philosophers of the time; Renè Descartes, Galileo Galilei, Robert Boyle, Thomas Hobbes, Henry More, and others, as well as absorbing the classics such as Euclid's *Elements of Geometry.* In these few short years he went as far as others could take him; then he went out on his own into uncharted waters.

Newton's notebooks reflect the same relentless curiosity and intellectual vigor as the famous ones of Leonardo da Vinci. But Leonardo's notebooks seem to be a haphazard accumulation of unrelated facts and random insights compared with Newton's, which were systematically organized to allow him to synthesize answers in a manner the great Leonardo never could.

However much one is impressed by Newton's rapid assimilation of the mathematical and physical literature in his first three years at Cambridge, what he accomplished in the next three years, justly called the *anni mirabiles*, is simply beyond comprehension. In the years 1664 through 1666, Newton invented calculus; developed his theory of light and color; became the first person to understand in the modern sense

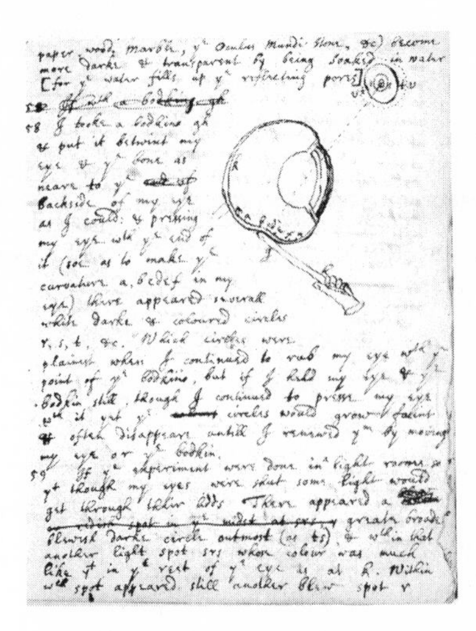

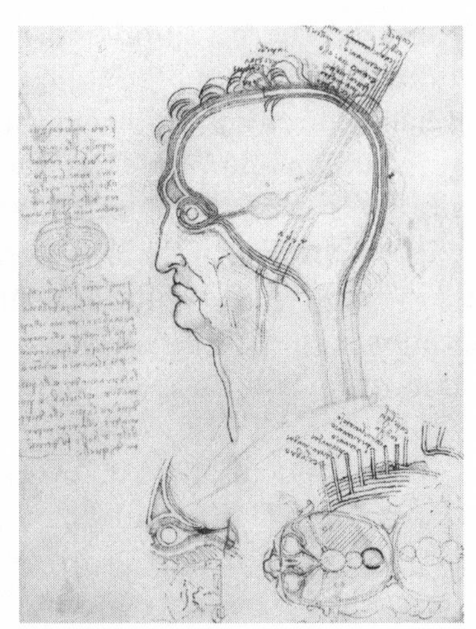

Pages from the notebook of Newton (left) and Leonardo (right). In the study of the eye Newton was not nearly the artist Leonardo was, but his notebooks have a sharper focus.

the concepts of force, inertia, velocity, acceleration (indeed, the complete science of dynamics); and had his first insights into the universality of gravity.

Newton's approach to natural philosophy was very much in the tradition of Galileo; it was grounded in experimental science. Nowhere is his dependence on experiment more apparent than in the development of his theory of light, colors, and optics. Not only did Newton perform experiments with prisms, lenses, mirrors, and the like, but he was fascinated with the optics of the human eye. Driven by intellectual curiosity, he stared at the sun so long in the course of his experiments that he almost lost his sight and had to stay in a dark room for several days to recover. And to discover how his vision would be changed if he modified the shape of his retina, he slipped a bodkin (a long, blunt sewing needle) "betwixt my eye & y^{e} bone as neare to y^{e} backside of my eye as I could." Again, he almost lost his sight.

Along with developing the science of optics, in the years 1664 through 1666 Newton's inquiries turned to dynamics and how it might be used to account for planetary motions. Kepler had provided a mathematical description of *how* the planets move around the Sun. The next step was to understand *why* the planets move around the Sun. This was a giant question that required answers to important preliminary questions: What is the nature of the forces responsible for celestial motions? What is the "natural state" of motion? What kind of force or impetus is responsible for the elliptical orbits of the planets? Can one understand the reason for Kepler's other two laws of planetary motion? The ultimate goal was finally to reunite physics and astronomy and to understand the solar system on the basis of physical laws. But before he could understand how forces could move planets, Newton had to develop a theory of how forces operate on Earth, even to decide exactly what is meant by a force. The state of affairs in 1664 when Newton turned to dynamics had just been described as an open book. But to appreciate Newton's accomplishment, one must to scan the contradictory and fragmentary information others had written in the opening pages of the book of dynamics.

In the first pages of this book were entries by Aristotle, Kepler, Galileo, Gilbert, and Descartes, as well as others who can hardly be described as slouches. When we read the opening pages of the development of dynamics, we do not find a slow, smooth march of science, but a confused dance of bodies, sometimes moving in straight, orderly lines, sometimes traveling in circles, often going backward, with objects sometimes moving by their own impetus, other times dancing under the influence of unseen forces, or even being dragged behind or pushed along by whirlpools in an aether. The dance of dynamics included steps (and missteps) by the following:

ARISTOTLE: Aristotle held that the universe was divided into two parts, the terrestrial region and the celestial region. In the realm of Earth, all bodies were made out of combinations of four substances, *earth, fire, air,* and *water,*[5] whereas in the region of the universe beyond the Moon the heavenly bodies such as the Sun, the stars, and the planets were made of a fifth substance, called *quintessence.* Heavy material bodies like rocks and iron consisted mostly of *earth* with small parts of the other elements. Less dense objects were thought to contain a larger admixture of the other elements along with *earth.* For instance, humans

[5] Here the elements are denoted by italics. Thus, *earth* is a pure element, whereas Earth is a planet made mostly of *earth* but also containing some of the other elements; *air* is a pure element, whereas the air we breathe is mostly *air,* but with some other elements mixed in.

consisted of a complex mixture of all the elements: *earth,* which gave material strength and weight; *fire,* which provided warmth; *water,* which accounted for blood and other bodily fluids; and *air,* which filled the lungs and provided the breath of life. Of course, some people were more earthly, fiery, airy, or watery than others. The Sun, planets, and stars were made of *quintessence,* a pure, perfect substance, quite unlike the elements found on Earth. The Moon, marking the boundary between the sublunary earthly region and the supralunary heavenly region, was mostly *quintessence,* but because of its proximity to Earth it was contaminated with a small admixture of earthly elements, which accounted for the visible imperfections on its surface.

The fundamental assumption in Aristotelian physics was that the natural state of sublunary matter is rest. *Earth, air,* and *water* must seek their natural place at rest in the center of Earth unless stopped by an impenetrable surface like the ground or a table. The natural place of rest of the element *fire* is somewhere above us (but well below the Moon). The air we see around us is a mixture of the elements *air* and *fire* (after all, air, at least in Greece, has warmth), so its behavior is complicated by the competition between the tendency for *fire* to rise and *air* to fall. Except in very complicated situations such as when *air* and *fire* were mixed together, motion was not a natural state of affairs.

Aristotle's model provided a simple, compelling explanation for falling rocks, rising flames, and the circulation of the air. However, it was less successful in explaining "violent motion" such as when an object is hurled from a catapult. To see why this would be a problem for the Aristotelian worldview, imagine the following experiment: Find a cat, and pult it from a siege machine. You would observe that the cat continues to travel through the air (before landing safely on its feet) even after it was no longer being pushed by the arm of the machine. If the natural state of motion of the cat is rest on Earth, why didn't the cat drop to the ground immediately on leaving the pult? Here, Aristotelian physics had to say that this kind of motion is different because it is "violent," and had to invent some mechanism to keep the cat in the air during violent motion. All of the mechanisms fall under the technical description "hand waving." One of the most popular explanations was that the air in front of the cat became disturbed by the movement of the cat, and swirled behind the cat and pushed it along. Thus, in Aristotelian dynamics, there was a distinction between "natural" downward motion (for example, a rock falling to the ground when dropped) and

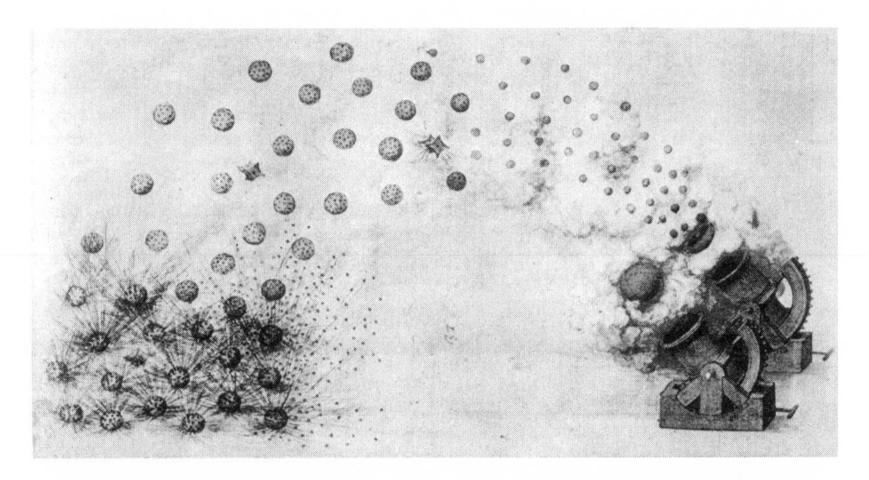

Leonardo was also interested in violent motion of projectiles, but mostly in his role as a military engineer. These drawings were for cannon design rather than understanding the motion of the projectiles. But clearly Leonardo knew that the projectiles travel in parabolic paths, and not according to Aristotle.

"unnatural" violent motion not directed toward the center of Earth (such as that resulting from a catapult).

In contrast to earthly motions, in the supralunary regions of the heavens the natural state of motion was circular, because circles were considered to be the perfect geometric figure. Thus the planets would travel forever in circular orbits without the intervention of any force or impetus, because, well, it's the natural thing for planets to do.

Although there was some degree of experience and observation in the physics of Aristotle, at its heart was a philosophical approach to science where the laws of nature are constructed to conform to a particular philosophical outlook. This basis for the investigation of nature led to some strange statements by Aristotle—for instance, that women have fewer teeth than men. Either Aristotle was not a very accurate observer, he couldn't count, or he had odd taste in women.

Although Aristotle had been the first and last word on dynamics for two millennia, after the work of Galileo, Descartes, and others it had become clear to the leading natural philosophers that a new system was needed. Although in the intellectual vanguard of physicists the feeling was that the physics of Aristotle was dead, in the curriculum of many universities in the mid-seventeenth century Aristotelian physics was not yet buried. Now we think of universities as the petri dishes of society where all sorts of new ideas and philosophies are grown (and, thankfully, most are discarded). But in the sixteenth and seventeenth centuries the universities were reactionary institutions dominated by Aristotelians, and they were not about to relinquish their authority without a struggle.

It is important to note that Newton's masterpiece was titled *Mathematical Principles of Natural Philosophy.* Here, by natural philosophy Newton meant physics, and by the title he made it clear that he believed that mathematics is the basis for physics, not philosophy, and in fact philosophy must be based on the laws of nature, as discovered by observation and understood in the language of mathematics. Nowhere is there a clearer reversal of the Aristotelian approach of basing the laws of nature on philosophy.

KEPLER: In his existence in a superposition of a modern person and an ancient person, Kepler's view of the forces responsible for the motions of the planets was a curious mixture of Aristotelian ideas and modern physics. Recognizing that the Sun was the true center of the solar system, he thought there must be some force emanating from the Sun that kept the planets in their orbits. However, he adopted the Aristotelian version of inertia that without a force acting on them, the planets would come to rest. Therefore he imagined the force of gravity to be a kind of sweeping force that constantly pushed the planets along in their orbits, and without the constant action of this force the planets (including Earth) would slowly wind down and come to rest like the hands of a mechanical clock that someone neglected to wind.

Kepler even hit on the right dependence of the strength of the gravitational force on distance. We know that the Newtonian gravitational force is an " **inverse-square**" force; that is, it decreases in proportion to the square of the distance between the objects. Kepler knew that the measured intensity of light also decreases as the square of the distance between the source of the light and the detector, and he reasoned that gravity would obey the same law. He was right that there is a deep connection between the reasons for the inverse-square laws for gravity and light intensity, but this seems nothing more than an educated guess.[6]

[6] Kepler reasoned that since the outer planets traveled more slowly about the Sun (his third law), the gravitational force must weaken with distance.

GALILEO: Galileo was perhaps the first person to grasp the modern concept of inertia: that a body set in motion will remain in motion with a constant velocity, and that motion is every bit as natural as rest. This was a tremendous leap of intuition, for in our experience everything does indeed sooner or later come to rest. Galileo had to turn the problem inside out and realize that the important question is not what makes things go but what makes things stop. To come up with the concept of inertia Galileo had to construct in his mind an unnatural world, with no friction and no air resistance, and to pose the problem in this imaginary world, not in the world of everyday experience. He also saw clearly that there was no difference between normal motion and violent motion.

But Galileo had the (to us seemingly funny) idea of circular inertia. He believed that in the absence of a direct force, planets would circle the Sun indefinitely, and somehow this was a result of, or at least related to, the fact that the Sun and planets rotate on their axes. This belief in circular inertia is one reason Galileo never accepted Kepler's law that the planets move in elliptical orbits. Galileo was someone who saw the big picture, and not a person to get lost in the details and worry about minor discrepancies like eight measly minutes of arc; and he was also someone who really appreciated only his own work. However, the true reason Galileo did not believe in elliptical orbits is that they did not fit into his picture of the physics behind the planetary motions.

Whether or not we realize it, it is impossible to form a cosmological model without the perspective provided by the laws of physics. In fact, a cosmological model is only as good as the physics on which it is based. It is equally true that a framework for understanding the laws of physics cannot be developed without a cosmological background, for cosmology provides the canvas on which we paint the picture of physics.

GILBERT: A possible connection between gravity and magnetism has attracted the attention of scientists from Kepler to Einstein, and is pursued even today. By Newton's time magnetic phenomena had been studied for centuries. One of the earliest surviving works discussing magnetic forces is the 1269 *Letter of Petrus Peregrinus on the Magnet.* Petrus Peregrinus (Peter the Pilgrim) had signed on with the forces of King Charles of Anjou for a crusade. In 1269 the forces assembled in Italy and felt they needed a little practice before starting the real crusade, so as a warm-up exercise they decided to lay siege to the Italian town of Lucera. Not every day during a siege is exciting, so Petrus had some time on his hands and out of boredom began toying around with lodestone, the mineral now known as magnetite, which is a natural magnet. Petrus discovered that all magnets have two poles and that lodestone can attract or repel iron or other bits of lodestone.[7]

In 1600, the Englishman William Gilbert (1544–1603) published his book on magnetism, *On the Lodestone and Magnetic Bodies*, in which he proposed that Earth is like a giant magnet, and the magnetic force is transmitted through some invisible magnetic effluvia. Gilbert reasoned that since Earth is a magnet, other planets, the Sun as well, should also have magnetic fields. He proposed that these magnetic forces were responsible for planetary motions. Although Gilbert's work is an important development in the study of magnetism, it only confused the issues in solar system dynamics.

DESCARTES: One problem Newton faced in putting together the pieces of the puzzle of planetary motion was that not only were pieces of the puzzle missing, but pieces that didn't belong to the puzzle were scattered on the board. Contributing yet another piece that didn't belong was the French mathematician and philosopher Renè Descartes (1596–1650). Although he ultimately rejected Descartes's central ideas of dynamics, Newton was greatly influenced by his philosophy and his development of analytic geometry.

Descartes correctly understood the modern idea of inertia, rejected the idea of circular inertia, and reasoned that the natural motion of a planet would be in a straight line with constant velocity unless acted on by some external agent. However, Descartes could not come to grips with the idea that gravity played a role in planetary motion because he could not accept the idea that a force could reach out and act on another body at a distance across empty space. Because he rejected the concept of **action at a distance**, he had to assume that the universe contained an aether filled with vortices that whirled the planets around. Of course, Descartes's vortex theory cannot account for Kepler's laws, and is pure speculation unsupported by any facts.

SUPPORTING CAST: As Newton looked through the first pages of the book, he could find many other proposals to explain planetary motion—some absurd, some clever, some reasonable, all wrong. It

[7] Much of what Petrus "discovered" was known to the Chinese at least as early as A.D. 1000.

Renè Descartes (left) and William Gilbert (right).

seems that every mathematician, physicist, and astronomer in Europe had a theory. Although they were from lesser luminaries, they too added to the confusion of the book's opening pages.

Typical of such models was one by the Italian astronomer Giovanni Alfonso Borelli (1608–1679).[8] In his model he assumed that planets have a natural tendency to approach the Sun, but their circular motion gives them a tendency to fly away from it, and the opposing forces counterbalance each other. Borelli postulated a relation between the orbital motion of the planets and the sun's rotation, and proposed that light rays from the Sun hit the planet and drag it along.

Borelli borrowed a little from everyone: from Aristotle, the idea that there are natural places of rest, and just as material bodies seek Earth, planets seek the Sun; from Copernicus, the view that the Sun is the center of the solar system; from Descartes, the point that inertia drives the planets out of orbit; from Galileo, the connection to rotation; from Kepler, the thought that somehow the force moving the planets was related to light rays. Put it all together, mix it up with a few vortices

[8] His book is about only the satellites of Jupiter, probably because after Galileo's troubles it was deemed wise to address questions about the motion of Earth obliquely.

if you like, and you have a hopelessly confused model incapable of any quantitative predictions. There were dozens and dozens of such models scribbled in the margins in the book of dynamics.

The purpose here is not to show how the heroes of science stumbled; indeed, the struggles of great scientists like Kepler, Galileo, and Descartes were simply an indication of the enormity of the problem now taught as a matter of fact in high school physics courses. The point is that the landscape Newton had to traverse was littered with the ideas of natural motions, violent motions, magnetic vapors, whirlpool vortices, repulsive gravity, attractive gravity, and circular inertia; lingering questions about the center of the universe; an incomplete understanding of the exact nature of inertia, momentum, weight, free fall; and no clear idea as to the motion of objects in space without the intervention of forces. Finally, a question that is irrelevant to modern physicists but that could not be ignored by seventeenth-century physicists: what is God's role in all this?

OUT FROM UNDER THE APPLE TREE

By 1664 Newton had read the first pages of the book of dynamics and seems to have been unintimidated by the expanse of blank paper awaiting him. During the years 1664 through 1666, Newton "more or less" worked out the theory we now know as Newtonian dynamics and applied it to planetary motions. Remarkably, it was not until 1687, with the publication of the *Principia*, that he shared the solution with the world. It is impossible to know to what extent in 1666 he understood (or believed in) what he had done. In his later years Newton intimated that indeed he had appreciated the enormity of his discovery as early as 1666. Furthermore, Newton implied that his tremendous achievement was simple for him. But if he truly understood it and believed in it, it is almost impossible to imagine that anyone, even someone as secretive as Newton, would wait twenty years before publishing it.

Everyone is familiar with the legend of Newton's "discovery" of gravity while sitting under an apple tree. According to the story, he observed an apple fall as he whiled away the day dreaming in his mother's garden. As the apple fell, the disparate pieces of the puzzle fell into place in Newton's head. Or so the story goes, for it was only in his later years that he told the story of the apple. No doubt the story originated with Newton himself, for there are four independent accounts of

Newton relating the story. In the version reported by John Conduitt (the husband of Newton's niece), the account goes as follows:

> ...whilst he [Newton] was musing in a garden it came into his thought that the power of gravity (wch brought an apple from the tree to the ground) was not limited to a certain distance from the earth but that this power must extend much farther than was usually thought. Why not as high as the Moon said he to himself & if so that must influence her motion & perhaps retain her in her orbit, whereupon he fell a calculating.

But in his later years Newton was quite concerned with adding to his legend, and it is quite likely that he polished the apple story a bit. It is impossible to know how much importance to attach to the apple.

The story of Newton's apple is perhaps the most aberrant tale in all of the legends of science. In many ways it trivializes Newton's accomplishments and gives an altogether false picture of the process of science. The universal theory of gravity and the science contained in the *Principia* did not come to Newton as a blinding insight while sitting under the tree. Looking through Newton's notebooks, one can trace the clear development of his ideas of forces, inertia, and gravity. He did not discover his laws of motion sitting under a tree, nor do I believe that he fully grasped the significance of his idea in 1666.

That dreamy discovery is the way many people think science is done is unfortunate. This is forcibly brought home to me in conversations with the general public, in particular with college students. One day, after a lecture on the foundation of Einstein's theory of relativity to a class of freshman nonscience majors at the University of Chicago, a student came up to me with a smile that seemed out of place the week before final exams. She said she thought the lecture that day was the best of the entire course. She went on to say that it had a very special meaning for her because she was from New Jersey, and once during a childhood visit to her grandmother in Princeton, her father had shown her the very tree Einstein was sitting under when he "dreamed up" his theory of relativity. The student hurried off to her next class before I could reply.

College professors are not usually at a loss for words, but this was one of those rare moments. I could never have imagined so many strange misconceptions packed into a single, harmless statement. First of all, she seemed proud to be from New Jersey. She also had the history wrong: Einstein went to the Institute for Advanced Study in Princeton in 1933 at the age of fifty-four, eighteen years after the publication of his general theory of relativity, which he finished at age thirty-six while a professor at the University of Berlin. Although the student hadn't mentioned a falling apple, obviously the apocryphal story of the seed of Newton's inspiration was mixed in.

But what really shocked me was the implicit assumption in her statement that science is conducted by following the simple steps of (1) finding a genius, (2) sitting the genius under a tree, and (3) waiting for the genius to have a great idea.

The true story of Einstein's development of his theory of gravity is much different. In 1907 Einstein was a happy man: not only had he recently been promoted from technical expert third class to technical expert second class at the Swiss patent office in Bern (with a welcome 29 percent increase in salary), but also he had what he called the "happiest thought of my life" when he postulated the principle of equivalence. In fact the equivalence principle, the simple idea that a freely falling person does not feel his own weight, was the subject of my lecture that day. But stating the principle was only the start of the process of discovery. After an incubation period of a few years, when Einstein made no progress on his theory, he doggedly pursued the implications of his principle. Over the intervening years he went down many blind alleys, and several times he thought he had the answer only to realize he had to start over. On more than a few occasions he wrote to friends in despair, ready to give up the struggle. The story of the development of Einstein's theory is one of dedication, determination, and no small measure of courage. It was a far cry from the scenario of a flash of divine inspiration coming to someone dreaming under a tree on a sunny day.

We know more of the process of the development of Einstein's theory of gravity than we know of Newton's. Newton was a secretive man, and it was only in his later years, when he was much concerned about his reputation, that he told the apple story.

We do know that Newton spent most of the time from 1665 to 1667 at home. From the summer of 1665 until the autumn of 1667, the plague visited the town of Cambridge, and the university was closed. During these years Newton returned to his family's estate in Woolsthorpe. We cannot reconstruct the path traveled by Newton's mind, for he was a proud man and, unlike Kepler, he didn't leave behind a record of his mistakes. It was only some fifty years later that Newton wrote about the events during the plague years:

> And the same year [1665] I began to think of gravity extending to y[e] orb of the moon & (having found out how to estimate the force w[ch] [a] globe revolving within a sphere presses the surface of the sphere) from Kepler's rule of the periodical times of the planets being in sesquialterate [three-half power] proportion to their distance from the center of their Orbs, I deduced that the forces w[ch] keep the planets in their Orbs must [be] reciprocally as the squares of their distances from the centers about which they revolve: & thereby compared the force requisite to keep the Moon in her Orb with the force of gravity at the surface of the Earth, & found them *answer pretty nearly* [my emphasis]. All this was in the plague years 1665–1666. For in those days I was in the prime of my age for invention & minded Mathematicks & Philosophy [natural philosophy, that is, physics] more than at any time since.

It all sounds rather neat and tidy.

In fact, it sounds too neat and tidy to ring true. In particular, much has been made of the phrase *answer pretty nearly*, for the Newtonian theory of gravity does much better than answer pretty nearly; in fact, it is in spectacular agreement with Kepler's laws and observations. This has suggested to many people that Newton did not realize that what he had discovered was not an approximation, but at least to the degree that one could determine at the time, an exact description of nature.

Newton set about writing of his discovery only after being challenged to do so. In 1684 the famed architect of Saint Paul's Cathedral in London, Sir Christopher Wren, issued a challenge to the British scientist Robert Hooke and the astronomer Edmond Halley, with a small prize (a book worth forty shillings) to the first person who could prove that Kepler's first law of elliptical orbits resulted as a consequence of an inverse-square law of gravity (as proposed by Kepler himself, and later suspected by Halley, Wren, and Hooke).[9]

[9] Wren and Halley were supporters and friends (to the extent that he had friends) of Newton. Robert Hooke, on the other hand, was an implacable enemy. Newton and Hooke first clashed over priority disputes in the field of optics, and remained rivals until Hooke's death in 1703.

In August 1684 Halley visited Newton in Cambridge on other business and happened to mention the challenge. Halley was astonished when Newton replied that he had proven that nearly twenty years previously. Three months later Newton delivered to Halley a nine-page paper titled *De Motu Corporum in Gyrum (On the Motion of Bodies in Orbit).* On reading the manuscript a stunned Halley realized that Newton had gone far beyond the original question, and demonstrated that *all* of Kepler's laws were the result of an inverse-square force. Halley realized that the nine-page treatise contained more than just a simple step forward, but a revolutionary discovery, perhaps equal in magnitude to any in the history of science.

Sometime during this period Newton realized that things did better than "answer pretty nearly," and in fact, rather than just an approximation, the universal law of gravity was a profound and fundamental insight. Newton requested that Halley keep the manuscript secret, sealed in the archives of the Royal Society with the date recorded to establish his priority. He then set about to write a longer detailed manuscript. But now Newton glimpsed the enormity of the problem he had solved, and although Halley might have expected to receive a longer version of *De Motu* within a couple of months, Newton saw a vaster horizon.

On grasping the magnitude of the question, Newton fell completely under the grip of the problem. In the image of Kierkegaard, the truth had Newton in its grasp and pursued him relentlessly. As Newton's servant Humphrey Newton (no relation) noted of Isaac,

> So intent, so serious upon his studies, y^t he ate very sparingly, nay, ofttimes he has forgot to eat at all, so y^t going into his Chamber, I have found his Mess untouch'd of wch when I have reminded him would reply, Have I; & then making to the Table, would eat a bit or two standing.... At seldom Times when he design'd to dine in ye Hall, would turn to ye left hand, & go out into ye street, where making a stop, when he found his Mistake, would hastily turn back.

As Newton wrote to Halley, "to do this business right is a thing of far greater difficulty than I was aware of." Indeed, before Newton could finish the treatment to his satisfaction he had to perfect the science of dynamics.

In the years following Newton's first draft of *De Motu*, it expanded from its original nine pages into two books (which became Books II and

III of the *Principia*) of around one hundred pages, preceded by another treatise, *De Motu Corporum (On the Motion of Bodies)*, which became Book I of the *Principia* containing the Newtonian theory of dynamics.

With the publication of the *Principia* the sciences of mathematics, physics, and astronomy, which had been separate for so many millennia, were joined in their modern form.

The book was empty no longer.

THE SUBLIME HAND OF BEAUTY

Reactions to the publication of the *Principia* were varied. It certainly could not be ignored or dismissed. The response ranged from that of people who could not accept the idea of action at a distance (even some of the best scientists of the time, like Hooke, Huygens, and Leibniz) to that of people who regarded it as revealed truth and immediately saw it for the remarkable achievement it was. Typical of the response of the latter was that of a famous French mathematician, the Marquis de l'Hôpital, who, on being shown the *Principia* (as reported by Conduitt), "cried with admiration 'Good God what a fund of knowledge there is in that book?' He then asked the Dr. [John Arbuthnot, the person who presented the book to l'Hôpital] every particular about Sir Isaac even to the color of his hair and asked does he eat & drink & sleep, is he like other men?"

Is he like other men? That is not an easy question to answer, for indeed Newton was a man, but he was unlike any other. He ate and he drank and he slept (but not often when pursued by a problem). He was also like other men in that he had more than his share of very human frailties such as petty jealousies, arrogance, ingratitude, rudeness, and in his later life some degree of avarice; in general, he was not a pleasant fellow.

But in a very real way he was not like other men. Indeed, to many in his own time, and to many more in ours, the *Principia* seems more astounding than merely impressive. As Subrahmanyan Chandrasekhar, whose work in astrophysics merited the 1983 Nobel Prize in physics, expressed it, the creation of the *Principia* is "an incredible accomplishment...beyond human comprehension. This rapidity of execution, besides the monumental scale of the work, makes the achievement incomparable." To exaggerate the leap made by Newton is nearly impossible (but I will try). To put the flight of his imagination into the familiar twentieth-century context of powered flight, we might compare the relative advance of Newton over his contemporaries to the advance of the Wright brothers over theirs. People might have been impressed that the Wright brothers flew an airplane at Kitty Hawk, but to compare the accomplishment of Newton to that of the first manned flight, one would have to imagine Orville and Wilber Wright pulling up on the sands of Kitty Hawk on December 17, 1903, behind the controls of a modern jetliner and flying off to New York.

THE LONGEST, SLENDEREST THREADS

Newton's leap of insight was all the more remarkable when one considers that he traveled across a chasm where so many others had fallen, for central to the Newtonian picture of the solar system is that the force of gravity acts at a distance across "empty" space. Descartes found this idea implausible, and even after the publication of the *Principia*, both Huygens and Leibniz rejected the concept of "action at a distance," with statements that it is ludicrous to imagine that any force can act across empty space without a medium to transmit its effects.

Newton realized that he had crossed the chasm by swinging on the slenderest of threads, and was also quite troubled by the idea of action at a distance. In a letter to Richard Bentley, on February 25, 1692, he said:

> It is inconceivable, that inanimate brute matter should (without ye mediation of something else, wch is not material) operate upon & affect other matter wthout mutual contact.... And this is one reason, why I desired you would not ascribe innate gravity to me. That gravity should be innate inherent & essential to matter, so y^{t} one body may act upon another at a distance through a vacuum wthout the mediation of anything else by & through wch their action and force may be conveyed from one to another, is to me so great an absurdity, that I believe no man who has in philosophical matters a competent faculty of thinking, can ever fall into it. Gravity must be caused by an agent acting constantly according to certain laws; but whether this agent be material or immaterial, I have left to ye consideration of my readers.

It took a great amount of courage for Newton to propose a theory of the universe with the invisible hand of action at a distance its central tenet.

For someone to make an entry into the book of knowledge requires a certain degree of self-assuredness, because no one, not even Newton, knows that his or hers is the complete story.

Newton demonstrated that the forces responsible for the motion of the planets are capable of human comprehension, and that they can be studied in terrestrial laboratories. Thanks to Newton we know that the very same force that keeps our feet planted on Earth keeps the planets in orbit. The heavens are comprehensible to us, for nature does not employ different forces in different parts of the universe. As the twentieth century physicist Richard Feynman put it, "nature weaves her tapestries from the longest of threads."

With a single sweep of genius, Newton freed celestial dynamics from the impression that crystalline spheres, mechanical gears, and other sundry devices moved the planets. Is there anyone who can fail to appreciate the simplicity and beauty of the Newtonian view of the solar system? To those who say Newton removed the hand of God from the heavens, I say he replaced a toilsome hand of brute force with a sublime hand of beauty.

So finally, in 1687 with the publication of the *Principia*, mathematics, astronomy, and physics were fully reconciled (hopefully never to part again), and out of the union issued a nearly complete understanding of the motions in the solar system.

Now we turn to the stars.

PART II

The Galaxy

SIX

The Third Dimension

People lucky enough to have been more than about five years old in the summer of 1969 carry around in their memory the unforgettable images of Neil Armstrong and Buzz Aldrin walking around on the surface of the Moon. That a member of our species traveled to another heavenly body and returned was a spectacular triumph for humankind. Yet a surprisingly large percentage of people believed that the event never occurred, but was staged somewhere in the desert in the southwestern United States. That the percentage of disbelievers was about the same as the percentage of people who now believe in abductions by space aliens is probably not a coincidence.

One quiet skeptic in the summer of 1969 was Elnora McGee. The disbelief of many was grounded in the deep paranoid suspicion of government characteristic of our society, but Elnora was by nature a trusting soul. She didn't really think the government would lie to us intentionally; she just thought that it was physically impossible for someone to walk around on the Moon.

One day that July while I was watching the Apollo 11 mission on television, Elnora took a break from her household duties and quietly came into the room and stared first at the television and then at me. I could tell something was troubling her, so I asked what she thought of the Moon mission. Although I was sitting and she was standing, she looked me straight in the eye (she was all of 4'10") and asked,"Do you really believe those men on TV are going to walk on the Moon?" As a seventeen-year-old, recently graduated from high school, I knew just about everything, so I told her sure, if the landing module deploys successfully, if the retro-rockets fire properly, if the trajectory of the descending orbit is calculated correctly, if the guidance system works properly, and so on.

But even after hearing such a splendid answer exhibiting all the expert knowledge recently gained by reading an article in *Scientific American* magazine, she was still staring at me slowly shaking her head. She obviously appreciated the fact that since I was going to be a physics major in college I knew all sorts of things, so she asked me the question that was troubling her. She pointed toward the ceiling in the direction of the light coming from a bare bulb where just the week before had hung a chandelier until smashed by the impact of a golf club, and asked, "Do you mean to tell me that someone could just fly up to the ceiling and walk around on that light bulb?"

For many years I thought that it was just about both the funniest and the saddest thing I had ever been asked. I related the story to many friends, and we all chuckled that someone could believe that the Moon was just like a light bulb in the sky, hanging there a mere ten feet overhead. I also was saddened by what it implied about the educational opportunities available to a black woman born in 1903 in the Deep South. She might not have finished the third grade, let alone have taken a high school physics course.

Two and a half decades later, after struggling for many years trying to determine the size and distance of astronomical objects, I have come to appreciate that Elnora's question was quite erudite. After all, the ancient Mesopotamians apparently believed that if they could just find a high enough mountain or build a tall enough building, they would be able to reach up and touch the stars. In Tycho's time it was thought that supernovae were atmospheric phenomena, and Galileo himself believed that comets were closer to us than the Moon. How *do* we know the distance to the Moon, to the Sun, to the planets, and to the stars? Although some people may not know the actual distance to these things, they do *know* that they are "astronomically" far away. But how do they know that? They can't tell just by looking at them.

The determination of the distance to objects in the heavens has always been the fundamental problem in astronomy. We can chart the position of objects on the **celestial sphere** with great precision reasonably easily, but when it comes to determining how far away they are, we have to grope around and struggle pretty hard to find an answer. Knowing the apparent position of an object on the sky provides information about the location in the universe of the object in two dimensions (say, equivalent to its latitude and its longitude on the celestial sphere). But for a complete description of the universe, we have to know that third dimension, the distance to the object. Without knowledge of the distance to celestial objects, our view of the universe is without depth, rendering our picture without perspective, as flat and lifeless as paintings before Leon Battista Alberti's *On Painting* taught artists of the early fifteenth century the art of perspective.

Dogs might howl at the Moon without caring how far away it is, but our relentless curiosity about the natural world drives us to discover the distance to things we see in the sky. Perhaps our intellectual horizon in intertwined in a deep way with our perceived size of the universe. The exploration of the third dimension includes everything you might expect in a human intellectual endeavor: error, prejudice, stupidity, stubbornness, ingenuity, insight, intuition, and genius. The quest for perspective continues still, with many successes and some frustrations.

After Newton's time, cosmologists turned from the solar system to questions about the stars beyond. One of the most formidable tasks, before they could understand the stars, was to fathom the distance to them. Only by determining the distance to the stars can we explore the universe beyond our galaxy. But even before measuring the distance to the stars, first we must measure the size of the Earth, and the distances to the Moon and the Sun.

We can get a reasonable idea of the scale of distances to the Moon and the Sun without telescopes, fancy instruments, or complicated mathematics and physics. In fact, the determination of the distance to the farthest reaches of the universe starts rather modestly, with a hole in the ground.

EARTH, THE MOON, AND THE SUN

Distances are always measured in some units: inches, centimeters, miles, light-years, and so on; it really doesn't matter which you use. And to measure a distance between two points, you take a ruler marked in your favorite units and place it between the points. Or if that is inconvenient, you might hire a professional surveyor to look through a sighting scope at a companion standing around in a bright yellow vest waving a little red flag.

But if we can't reach out and stretch the ruler between us and a star, or even station a surveyor there with a little red flag, how can we hope to measure the distance to a star? In the sky, just as on Earth, unknown distances are always measured in terms of something smaller that is known. For instance, the distance to nearby stars is determined as some multiple of the distance between Earth and the Sun; the distance from Earth to the Sun and the Moon is determined in terms of the radius of Earth. So the first step in the determination of the distance to the stars is to measure the radius of Earth.

Noting that the shadow of Earth cast on the Moon during a lunar eclipse is always round, Aristotle realized that Earth was spherical. Like Christopher Columbus, the ancient Greeks knew that Earth was round, but unlike Columbus, their estimate of the radius of Earth was reasonably accurate. Around 200 B.C. the Greek astronomer Eratosthenes determined the radius of Earth to be about 3,750 miles, very close to the actual value of 3,960 miles.

The method Eratosthenes used was a relatively simple exercise in geometry. He knew that at noon on the date of the summer solstice the Sun was directly overhead the Egyptian city of Aswan, because the Sun shone directly down a deep hole in the ground on that day.[1] He also determined that on that day the Sun was not quite directly overhead in his home city of Alexandria, but was about 7° away from directly overhead (the zenith). The radius of Earth can then be determined in terms of the distance between Aswan and Alexandria, about five hundred miles.

Eratosthenes' determination of the radius of Earth was a beautiful illustration of the power of geometry, and gave a very accurate result. It is remarkable that in an age when even the most adventurous human had traveled not much more than about a thousand miles from home, the Greeks accepted the fact that the circumference of Earth was about twenty-five thousand miles.

Once the radius of Earth is known, the size and distance to the Moon can be determined, again by simple observations and the elegant

[1] Aswan is at about 24° north latitude, so when the Sun reaches its highest point in the sky on the summer solstice it is very nearly directly overhead.

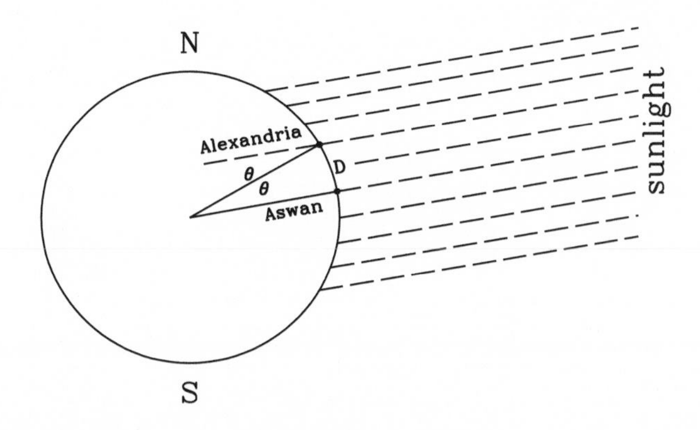

Eratosthenes' determination of the radius of Earth: When the Sun was directly overhead in Aswan, it was at an angle θ of about 7° from the zenith in Alexandria. If the distance between Alexandria and Aswan is known, the radius of Earth can be calculated in terms of that distance and the angle θ.

application of geometry. As Elnora McGee astutely realized, we can't find the true physical size of the Moon just by looking at it. The only thing we can see is that the Moon has an **angular size** of about one-half of a degree. It could be close to us and the size of a light bulb, or very far away and large enough for Neil Armstrong to walk on. But with a little work we can resolve the ambiguity.

One particular method was pioneered about 250 B.C. by the Greek astronomer Aristarchus (the same person who proposed that the Sun was the center of the solar system). Aristarchus could see that the angular size of the Sun is also about the same as that of the Moon, about one-half of a degree, because they both appear to be the same size.[2] If the Sun is much larger than Earth, then the shadow cast by Earth also forms an angle of about one-half of a degree.

During a lunar eclipse the Moon falls in the shadow of Earth. Aristarchus noticed that during a lunar eclipse the shadow of Earth seems to cover an area larger than the Moon. He estimated that the diameter of the Moon is about three-eighths as large as the shadow cast by Earth.

[2] Because the Sun and Moon are about the same angular diameter, during a *solar* eclipse the Moon just about exactly covers the Sun.

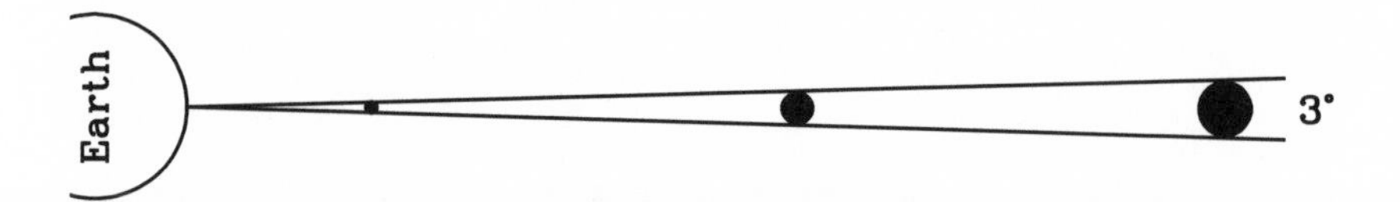

Elnora's dilemma: We directly observe the angular size of the Moon, not its true size. This leads to an ambiguity in the true size of the Moon. It can be small and close by or large and far away and still have the same angular size. For the picture above, the angular size of all three objects is 3°. (The angular size of the Moon is smaller, about one-half degree.)

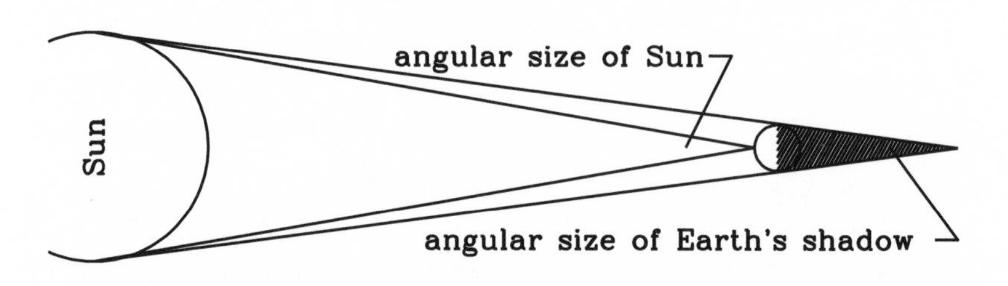

The Sun's shadow: If the Sun is much larger than Earth, then the angle formed at the point of Earth's shadow is about equal to the angular size of the Sun viewed from Earth, about one-half degree. (The angles shown in the diagram are much larger than one-half degree.)

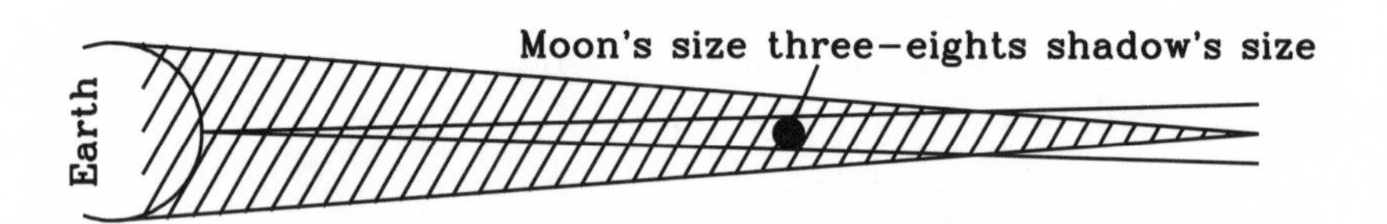

Aristarchus's determination of the size and distance to the Moon: Knowing the angle formed by Earth's shadow and the angular size of the Moon, there is only one place to put the Moon where it covers three-eighths of the area of the shadow. This fixes the actual size and distance to the Moon in terms of Earth's radius.

Simple geometric considerations, along with these observational facts, led Aristarchus to conclude that the Moon was one-fourth the size of Earth (very nearly the correct value), and the distance to the Moon was about sixty times the radius of Earth, again very close to the correct value.

Even without a detailed understanding of the numbers and geometrical arguments, anyone can appreciate the power of simple geometry coupled with reasonably crude observations. The determination of the true size of the Moon and its distance from us does not require fancy instruments or telescopes, or complicated physics, but along with a rudimentary knowledge of geometry, only a hole in the ground to determine the radius of Earth and the patience to watch the sky during a lunar eclipse. Elnora could have done it. If I had appreciated her question, I could have told her how.

It is no accident that the size of and distance to the Moon is determined in terms of the radius of Earth, for we always measure unknown distances in terms of a smaller known distance. We will determine the distance to the far reaches of the universe by first determining the radius of Earth, using that knowledge to determine the distance to the Sun, using that to determine the distance to nearby stars, using what we learn about nearby stars to determine the distance to even more distant stars, and so on. The journey to distant horizons is taken not in a single bound but in a series of small steps, with each advance the result of determination, insight, and sometimes damn hard work.

The most complete model of the solar system in antiquity was developed by Ptolemy in the second century A.D., and it would serve as the standard model of the universe for thirteen hundred years. By the time Ptolemy constructed his cosmological model, he had reasonably accurate information about the size of Earth and the Moon, and the distance from Earth to the Moon. Of course, Ptolemy needed more than this for a cosmology. The most useful unit of distance in the solar system is the Earth–Sun distance, but there was no simple, reliable way for him to obtain this. The best estimate was based on **Aristarchus's method for determining the Earth–Sun distance**. The method is based on the fact that the time between the new moon and the first quarter of the moon is slightly shorter than the time between the full moon and the last quarter. Again, using the elegance of geometry, Aristarchus was able to express the Earth–Sun distance in terms of the time difference between these two intervals, and the Moon–Earth distance (which in turn was known in terms of Earth's radius, which was known from the hole in the ground in Aswan).

OBJECT	DISTANCE FROM EARTH (IN MILES)		RADIUS (IN MILES)		ANGULAR SIZE (IN DEGREES)	
	Ptolemy	True	Ptolemy	True	Ptolemy	True
EARTH	—	—	*3,750*	*3,960*	—	—
MOON	*225,000*	*239,000*	*940*	*1,080*	*1/2*	*1/2*
SUN	*4,300,000*	*93,000,000*	*21,000*	*433,000*	*1/2*	*1/2*

Sizes and distances of Earth, the Moon, and the Sun used by Ptolemy compared with their actual values.

Today we know that this time difference is about a half hour, but without reliable clocks, it was impossible for Aristarchus to measure this small of a time difference over the course of a month. Aristarchus greatly over estimated the time difference and thought it was twelve hours, rather than a half hour. Although his geometry was perfect, this error led him to conclude that the Earth–Sun distance was 1,260 Earth radii, or about 4.7 million miles, whereas the true distance to the Sun is about 23,000 Earth radii, or 93 million miles.

Once the distance to the Sun is known, as well as its angular size, the true physical size of the Sun can be calculated easily. Using the incorrect value for the distance to the Sun led Ptolemy to believe that the Sun was five and one-half times larger than Earth (the correct ratio is 110).

Although Ptolemy underestimated the Earth–Sun distance and the size of the Sun, his numbers were still enormous by human standards. He also knew that the Sun was quite a bit larger than Earth.

Ptolemy was well aware that the weak link in the distance chain was the determination of the Earth–Sun distance, so he expressed all distances in his solar system model in terms of this fundamental (but poorly known) distance.

THE PLANETS

The distances and sizes of Earth, the Moon, and the Sun could be found without reference to any cosmological model. It really didn't matter whether the Sun goes around Earth or Earth goes around the Sun. But to determine the distances to the planets, it is necessary to work within the context of a model for the arrangement of the solar system.

Ptolemy's model, roughly based on Aristotle's cosmological worldview, is that Earth is the center of the solar system, and at rest, with each planet orbiting Earth. The two major components of each planetary orbit are spheres, called the **deferent** and the **epicycle**.[3] The deferent is a large sphere surrounding Earth which carries on it another smaller sphere, the epicycle, which in turn carries the planet.

By fitting the observed motions of the planets, Ptolemy was able to determine the *relative* size of the epicycle in terms of the size of the deferent. But each planet had its own epicycle and deferent, and to fit the

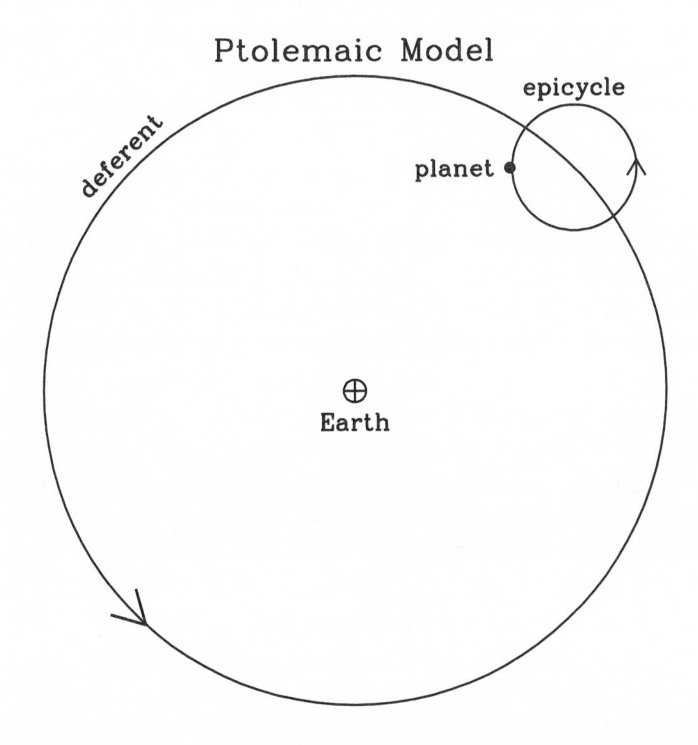

Ptolemy's model: Each planet rotates on a sphere known as the epicycle, which is carried around Earth on a sphere known as the deferent. To understand this geocentric model in heliocentric terms, the deferent is related to the planet's motion about the Sun, and the epicycle is related to Earth's motion about the Sun, which causes an apparent displacement of the apparent position of the planet as viewed from the moving Earth.

[3] This is a somewhat simplified version of Ptolemy's model. Some of the baroque features of the model are discussed in "The Devil in the Details" in this book.

pieces together and find the true sizes of the system he had to make two additional assumptions.

The first additional assumption was that the order of the planets going out from Earth corresponded to their apparent periods of revolution around Earth. With this assumption the order of planets was the Moon, followed by Mercury, Venus, Sun, Mars, Jupiter, and Saturn (in Ptolemy's model the Sun and the Moon were considered planets, and of course Uranus, Neptune, and Pluto cannot be seen without the aid of a telescope).

The first assumption set the order of the planets, but he still had no idea of their distances from Earth. Here a second assumption was necessary. Probably motivated by a perceived need for a mechanism to move them, Ptolemy imagined that the spheres of the planets were tightly nested, with no space between the spheres carrying the orbits of the planets. This meant that the greatest distance of one planet from Earth is the same as the least distance of the next most distant planet. Thus, if he knew the distance to any planet (other than the Moon, which was the innermost planet in his model), he would be able to nest the planets together and fix the distance to all of them and have a complete picture of the solar system. Since the "planet" he knew the distance to was the Sun, all dimensions in his planetary model were based on the Earth–Sun distance. Because he underestimated the Earth–Sun distance by about twenty, all distances in his model were too small.

Once the distance to a planet is known, its physical size can be determined if one can measure its apparent angular size. To do this, one must to *resolve* the image of the planet and see that it is not just a "point" source of light.

The planet with the largest angular size is Venus. It comes relatively close to Earth and is not too much smaller than Earth. The planet with the next largest angular diameter is Jupiter. Although it is much farther away than Mercury or Mars, its physical size is so much larger than the relatively small Mercury and Mars that its angular size is larger. The angular size of Venus is about one-half minute, and the angular size of Jupiter is about four-tenths of a minute. The human eye is capable of an angular resolution of slightly better than one minute of arc. Anything with a smaller angular size cannot be resolved and will appear as a point. Because the resolution of the human eye is about one minute of arc, it is impossible to resolve the planets without using binoculars or telescopes.

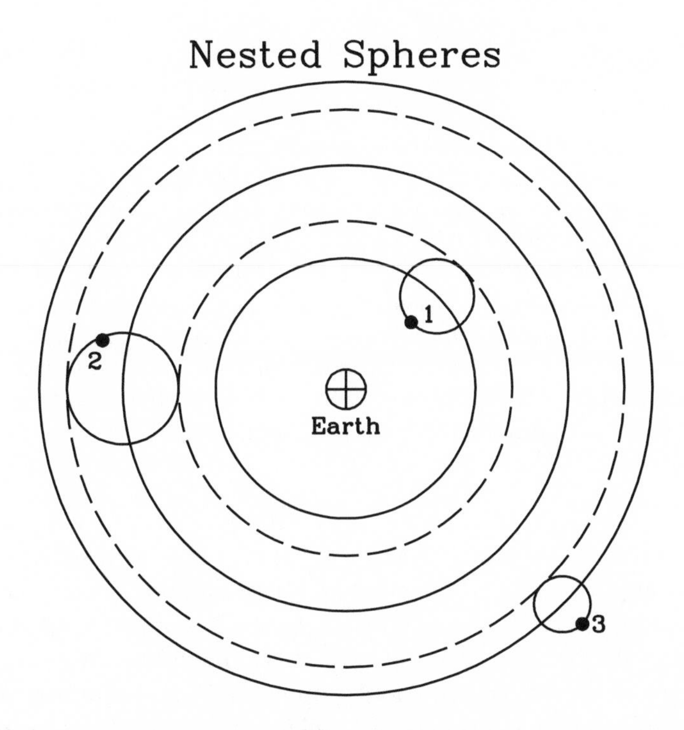

The nested spheres: Ptolemy assumed that the planetary spheres were nested, with no empty space between the greatest distance of one planet and the smallest distance of the next outermost planet. For instance, the closest distance of planet number 2 as it travels on its epicycle and deferent is the farthest distance of planet number 1. Similarly, the farthest distance of planet number 2 is the closest distance of planet number 3.

But no one told Ptolemy! Ptolemy believed that he could resolve the planets and tell that they were more than just "points" of light. He thought that he could detect a finite size for Venus of 3 minutes of arc, 2.5 minutes for Jupiter, 2 minutes for Mercury, 1.7 minutes for Saturn, and 1.5 minutes for Mars. These fictitious angular sizes for the planets were accepted without question until the invention of the telescope some thirteen hundred years later!

Although Ptolemy's values for the angular sizes of the planets are all wrong, it is certainly understandable how he arrived at them. Scattering of light in the atmosphere does tend to spread out the light from the planet leading to a seemingly larger angular size. But the real problem

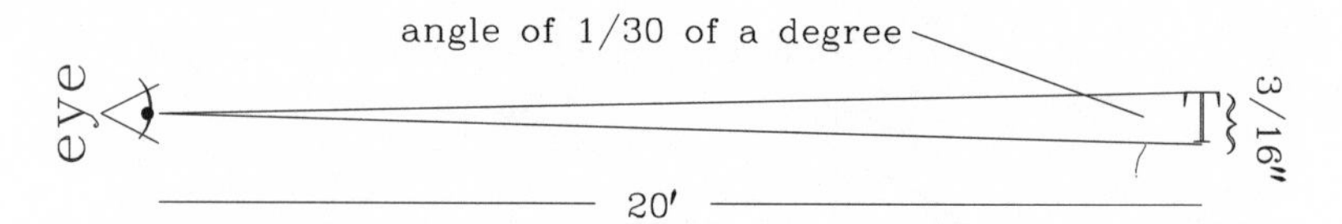

The resolution of the human eye: A person with normal 20-20 vision is capable of reading letters 3/16" high at a distance of 20' (the figure is not to scale—the angle shown is about 3° or about ninety times too large). This corresponds to an angular resolution of about two minutes of arc, or about one-thirtieth of a degree. The resolving power of the eye is slightly better than one minute of arc, which means it can see something about 1/16" high at a distance of 20'. Although a normal eye could tell something was there, it couldn't determine if it was an O, an F, an L, a C, or a T.

Ptolemy had is that he was working at the very limit of his instrument, in this case his eyes. He knew that he was seeing things that were just about as small as he could resolve. It is always a problem in astronomy, indeed in all of science, that the most interesting things are seen just at the limits of the power of the tools employed. Looking through a telescope designed to see things one million light-years away, you will invariably see hints of exciting things that seem to be one million and ten light-years away. You never really know whether or not what you are seeing is fictitious. There is always the demand to push things as hard as we can, and we are so easily deluded that we can see things that are not there. That is why scientists are always skeptical of any result that is "on the hairy edge" of detection.

Using the (incorrect) values for the angular diameters of the planets, Ptolemy could easily calculate their sizes. Of course, the sizes he deduced for the planets ultimately rested on the size of Earth, which we learned from the hole in the ground.

The planetary model of Ptolemy had several obvious flaws: the arrangement was wrong, the motions were wrong, the distances to everything but the Moon were too small, and the angular sizes of the planets were illusory. However, Ptolemy's approach was in every sense scientific. He worked within the confines of a model, made assumptions, measured all the parameters of the model that he could, and knew that if his assumptions were wrong, the results were wrong. In particular he realized that although the assumption of a stationary

PLANET	DISTANCE FROM EARTH (IN MILLIONS OF MILES)		ANGULAR DIAMETER (IN MINUTES)		ACTUAL DIAMETER (IN MILES)	
	Ptolemy	True	Ptolemy	True	Ptolemy	True
EARTH	—	—	—	—	*7,500*	*7,900*
MERCURY	*0.6*	*57*	*2*	*0.01*	*300*	*3,100*
VENUS	*4*	*26*	*3*	*0.5*	*1,900*	*7,700*
MARS	*33*	*49*	*1.5*	*0.15*	*8,600*	*4,200*
JUPITER	*53*	*391*	*2.5*	*0.4*	*32,500*	*89,000*
SATURN	*74*	*794*	*1.7*	*0.2*	*32,000*	*75,000*

Distances from Earth, angular sizes, and actual sizes of the planets according to Ptolemy and the modern values. In Ptolemy's model the distances are the maximum *distances between Earth and the other planets, and the true distances are the* minimum *distances.*

Earth was based on sense, and the assumption of the order of the planets out from Earth was based on the measured orbital periods, there was no physical justification for the nested sphere assumption and there could be arbitrarily large gaps between the planets. The best argument he could muster was the old "nature abhors a vacuum" cliché, writing that "it is not conceivable that there be in nature a vacuum, or any meaningless and useless thing." That there was an underlying physical reason for the nested spheres also seemed plausible. It is easier to imagine the celestial spheres rolling around on something and touching each other than to think that they were separated by the vacuum of empty space.

It is also remarkable that although his solar system was substantially smaller than we know it to be, it was no small place. To appreciate how large it appeared to him we must realize that it was no easier for the mind of antiquity to conceive of the seventy-four million miles to Saturn as it is for the modern mind to comprehend the billions of light-years used to describe the universe today. It is also interesting that although Earth in Ptolemy's model was the center of the universe, it was not the largest thing in it. Ptolemy believed (incorrectly) that Mars was slightly larger than Earth, and that the Sun was five and one-half times the size of Earth, while Jupiter and Saturn were more than four times larger than Earth.

✦ ✦ ✦

Ptolemy's nested-sphere model and his estimates for planetary distances and sizes were largely unquestioned until the acceptance of the Copernican arrangement. Making the assumption that the Moon orbits Earth, while Earth and the other planets orbit the Sun, Copernicus could estimate the distances to the planets in terms of the Earth–Sun distance. Just as in Ptolemy's model, the crucial unit is the Earth–Sun distance. Copernicus recalculated the distance to the Sun using the same methods as Ptolemy and arrived at a worse answer, some 6 percent smaller than Ptolemy's result.

In fact, the solar system was *smaller* in Copernicus's model. The greatest distance between Saturn and Earth in Copernicus' model was about forty-three million miles, nearly 30 percent smaller than Ptolemy's value. Although Copernicus stretched our human horizon, he didn't do it by proposing a larger solar system. However, Copernicus did obtain the proper arrangement of the solar system, and the model was reasonably accurate in terms of the relative distances of the planets from the Sun. In terms of the Earth–Sun distance, he found the distances from the Sun to be 0.36 : 0.72 : 1 : 1.5 : 5 : 9 for Mercury, Venus, Earth, Mars, Jupiter, and Saturn, respectively. (The true values are 0.387 : 0.723 : 1 : 1.5 : 5.2 : 9.5.) The *relative* scale of Copernicus's solar system was correct, but every distance hinged on the Earth–Sun distance, which was far too small.

Not until Giovanni Domenico Cassini (1625–1712) could use accurate measurements of the positions of Mars to determine the parallax of the Sun, and arrive at an Earth–Sun distance of only 7 percent smaller than its true value, was the size of the solar system finally established.

THE SHIFTING STARS

Many people expect that because I am a professor of astronomy and astrophysics I can recognize all the constellations and pick out planets in the night sky. Actually, I can identify a couple of constellations, but I wouldn't want to compete with a dedicated amateur astronomer in a game of *find the planets.* In my defense, anyone who has looked at the sky with the unaided eye can tell you that planets and stars really do look a lot alike. Ancient astronomers did not differentiate between stars and planets because of their appearance or their twinkling, but rather because of the peculiar motion of planets across the heavens. But since planetary motions cannot be discerned with only a quick glance at the sky, on any given night stars do look a lot like planets. It is understandable that ancient astronomers thought that planets and stars were made of the same stuff, and were about as distant from us. Perhaps my inability to pick out planets can also be forgiven.

Although speculation about the nature of stars and the distances to them was a popular philosophical pastime, really smart scientists evaded the question. One of the most talented evaders was Galileo Galilei. In *Dialog Concerning the Two Chief World Systems*, Galileo's mouthpiece Salviati, proponent of the new Copernican cosmology, asks his opponent Simplicio, the defender of the canonical Ptolemaic cosmology:

> Now what shall we do, Simplicio, with the fixed stars? Do we want to sprinkle them through the immense abyss of the universe, at various distances from any predetermined point, or place them on a spherical surface extending around a center of their own so that each of them will be the same distance from that center?

Of course, the spineless Simplicio doesn't have a strong opinion and waffles:

> I had rather take a middle course, and assign to them an orb described around a definite center and included between two spherical surfaces—a very distant concave one, and another closer and convex, between which are placed at various altitudes the innumerable host of stars. This might be called the universal sphere, containing within it the spheres of the planets which we have already designated.

Here Galileo presents a wonderful parody of someone who simply repeats what he has been taught without any real understanding. The words spoken by Simplicio are nothing more than a ritualistic incantation of Ptolemy's cosmology.

In Ptolemy's nested-sphere model, the stars were contained on a sphere just outside the orbit of Saturn, at a distance of about seventy-five million miles from Earth. This is no meager distance, but it is much less than the true distance to the nearest star, twenty-five million-million miles.

Ptolemy also estimated the size of stars. He incorrectly thought that he could just barely resolve the brightest stars, assigning to them an angular diameter of one and one-half minutes of arc. At the distance of seventy-five million miles, this would give the brightest stars a physical size of four and one-half times the size of Earth, or about the size that Ptolemy presumed for the Sun, Jupiter, and Saturn. Fainter stars appeared to have a smaller angular diameter, and since all stars were thought to be the same distance from Earth, it was assumed that the fainter stars were smaller than the brighter stars. Ptolemy's theory for the stars is plausible, but of course completely wrong.

Galileo cleverly sidesteps the question. The wonderful Italian word *furbo* may be applied to Galileo's treatment of the stars. The literal translation of *furbo* might be "foxy" or "sly," with the modern connotation usually in the complimentary sense. In fact, Galileo doesn't have a clue where to put the stars in his arrangement of the universe, but he won't admit it. So he turns the problem around by forcing his opponent to answer. He is willing to adopt as a working model whatever Simplicio suggests, for if his suggestion turns out to be wrong, Galileo can disown it as coming from Simplicio, and if it turns out to be correct, Galileo can claim it as his own. Of course, Galileo didn't know that both choices offered by the wise Salviati were wrong, and seems to forget to mention that the stars were considered a bit of an embarrassment in the Copernican cosmology.

In Copernicus's model, or any other model in which Earth moves, an additional ingredient allows a direct determination of the distance to the stars. The movement of Earth should lead to an apparent annual motion of the stars, known as the annual stellar **parallax**, or just parallax for short.

Parallax is the *apparent* change in the position of an object caused by a change of the viewing position of the observer. Like so many other physical phenomena, children understand it better than many adults. Every child has the experience of settling into a seat at the theater, only to have an adult sit in a seat directly ahead. Because it is not enjoyable to watch a movie with the back of someone's head in the middle of the screen, the child moves over a seat or two until the back of the tall person's head is no longer in front of the screen. By a change of viewing position, the position of the person's head is shifted with respect to the distant screen. (If the movie is really boring, you can calculate the distance to the person in front of you by measuring how far you moved, and estimating the angle by which the person's head appeared to move.)

The important lesson is that if your point of view changes, the relative positions of the things you see will also change. In the case of the theater, it was the relative position of the head and the screen. In Copernicus's model, or any other model that involved the motion of Earth, it would be the relative positions of nearby stars with respect to distant stars that would change as Earth orbits the Sun.[4] By measuring the angle by which stars seem to shift around over the year, one can determine the distance to the stars by the pure methods of geometry.

Perhaps the cleanest way to demonstrate that Earth moves is to detect the annual movement of the stars. However, the more distant the stars from us, the smaller the degree of their apparent motion. From the lack of any observations of yearly motions of the stars, Copernicus knew that the stars had to be much more distant than Ptolemy assumed. If the closest stars were at the distance of twenty-thousand Earth radii, as assumed by Ptolemy, for the Earth-Sun distance used by Copernicus the annual displacement of the closest stars would be about 6°. This would result in an enormous shifting about of the relative positions of the stars, with all of the constellations changing shape over the course of the year.

Copernicus realized the stars must be much farther away than assumed by Ptolemy, but how far away? If the displacements of the stars due to the motion of Earth were too small for detection by the naked eye, say less than a minute of arc, then the distance to the star would have to be more than 3.5 million Earth radii, or nearly two hundred times more distant than Ptolemy's estimate. Thus, while Copernicus slightly shrunk the size of the solar system, he greatly enlarged the distance to the stars.

Pushing the stars to such great distances seemed absurd to some of Copernicus's contemporaries. Although we tend to dismiss them as people of small vision, some astronomers opposed to the Copernican revolution were thoughtful people with a sound (although ultimately flawed) argument. For the stars in Copernicus's model to have the accepted apparent diameter of 1.5 minutes of arc, the physical radius of

[4] Of course, if all the stars were the same distance from us, as imagined on the ancient sphere of the heavens, they would all shift in the same manner, and no parallax would be detected. For a visible parallax, some stars must be closer than others.

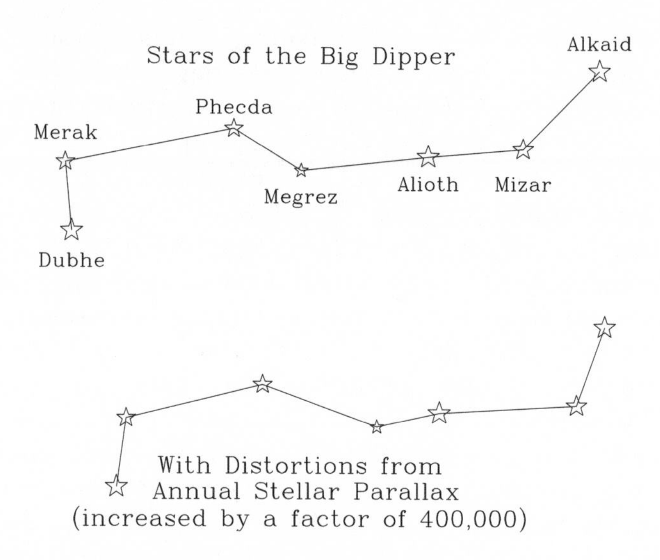

The stars of the Big Dipper change their relative position over the course of the year because of the motion of Earth, but the effect is too small to be noticed without telescopes. The star Megrez is the closest star in the constellation, about sixty light-years away, and Alcaid is the most distant, eight hundred light-years away. Viewed at six-month intervals, Megrez shifts during the year 0.05 seconds of arc, and Alcaid shifts 0.008 seconds. The shift of the other stars is between these two values. For the purpose of illustration, if we move in all the stars 400,000 times closer than they actually are to accentuate the effect, the parallax shift would be 400,000 times larger; 6° for Megrez and less than 1° for Alcaid. This would lead to an easily discernible distortion of the constellation over the course of the year as shown above.

the star would have to be larger than the distance between Earth and the Sun. This seeming absurdity led many reasonable people to reject the possibility of the movement of Earth. Of course, we know today that the angular size of stars is much, much smaller than believed in the sixteenth century, and in fact stars of such great size do exist. But we should still appreciate that Copernicus's model did not come without apparent absurdities, and reasonable people could find fault with it.

One might have thought that the great observer Tycho Brahe would have mounted a systematic search for stellar parallax to settle the question. However, there is no evidence that he took the possibility of a moving Earth seriously enough to waste his time searching. That is just as well, because stars are so distant that the annual parallax of the nearest star is less than a second of arc, sixty times too small to be noticed by the human eye, and the discovery of stellar parallax had to await the development of telescopes.

Although Kepler urged Galileo to search for stellar parallax with his new optic tubes, and Galileo must certainly have been aware that an annual parallax would be indisputable proof of Copernicus's model, stellar parallax is not mentioned in the *Dialog.* Either Galileo did not choose to undertake the arduous observation program required or he felt that his (incorrect) theory of the tides was proof enough that Earth moved.

As Copernicanism took hold in the 1640s and 1650s, nearly a century after the publication of *Die Revolutionibus*, the hunt for stellar parallax was on. The existence of an annual stellar parallax was a clear prediction of Copernicus's theory, and if it could be measured it not only would prove the motion of Earth but could be used to make the first direct determination of the distance to a star. With the reasonable assumption that the brighter stars are more likely to be closer than the dimmer stars, and hence display a larger parallax, astronomers started to search for any small yearly change in the relative positions of the brighter stars.

One of the first claims of detection of stellar parallax was by the English scientist Robert Hooke in the 1660s. Hooke was quite an interesting character: he considered himself the equal of Newton, with whom he constantly quarreled; he discovered many new phenomena in optics, mechanics, and astronomy; and he claimed also the discovery of just about everything else. He was a tireless self-promoter, whose motto might have been "I did it first." But in his defense, he did do many things first, and he was often right, occasionally spectacularly right.

Hooke claimed to have detected a parallax of the bright star gamma-Draconis. Most people doubted Hooke's value of fifteen seconds of arc for the star on the basis of only four observations over a period of only three months. Indeed, this was one of the times Hooke was wrong; the parallax of gamma-Draconis is about one hundred times smaller than Hooke's value. Hooke was wrong about the parallax, but he had stumbled onto something nearly as interesting.

In the 1720s James Bradley took a careful look at the motion of gamma-Draconis and discovered that any parallax of the star had to be smaller than one second of arc (indeed, it is 0.017 seconds of arc). However, Bradley realized that the apparent shift of the star Hooke had detected was a real effect due to the fact that light travels at a finite velocity, and the apparent direction of the incoming light ray must be corrected for the observer's motion. Although this effect, known as **stellar aberration**, proved the motion of Earth (as well as the fact that light travels with a finite velocity), by the 1720s everyone accepted the fact that Earth moved and by that time the motivation of the search for parallax was to measure the distance to the stars.

To imagine a second of arc, view this dot at a distance of one mile. At 10 miles the dot would have an angular size of 0.1 arc seconds; at 0.1 miles away its angular size would be 10 arc seconds.

It is also interesting that in addition to the discovery of aberration of starlight, the search for parallax led to two other spectacular serendipitous discoveries: the "proper" motion of stars on the celestial sphere by Edmund Halley in 1718 (the "fixed" stars aren't fixed after all) and the existence of double stars by William Herschel in 1803 (which proved that Newton's law of gravitation applied outside the solar system). Before one measures a parallax, one must take into account that stars have motions of their own (either the proper motion of stars as they randomly move about in the galaxy or the motion caused by the fact that some stars orbit about companion stars), and appear to shift position due to stellar aberration.

The first actual determinations of stellar parallax were made in the period 1836–1838 by two German astronomers, Friedrich Georg Wilhelm von Struve and Friedrich Wilhelm Bessel. Struve, who had emigrated to Russia to avoid military service, had been appointed director of the new Pulkovo Observatory in Saint Petersburg by Tsar Nicholas I, and Bessel was director of the Königsberg Observatory in Prussia. Both were talented astronomers who had the very best telescopes of the time at their disposal.

In 1837 Struve announced a determination of a parallax of 0.125 seconds of arc for the star Vega, remarkably close to the modern value of 0.123 seconds. Unfortunately, the tone of Struve's announcement was tentative, and even more unfortunate was his decision to remeasure the parallax—in 1839 he published a revised value of 0.262 seconds of arc, twice as large as the correct value. (The obvious lesson is always to be sure of yourself and never to check your results!)

The year after Struve's first result, Bessel determined a parallax of between 0.33 and 0.29 seconds of arc for the star 61-Cygni, the twelfth closest star to us, in good agreement with the modern value of 0.29 seconds. Bessel's measurement was widely accepted as the first unambiguous determination of stellar parallax, and the first measurement of the true distance to a star. A parallax of 0.29 seconds of arc implies that the star is at a distance of 700,000 times the Earth–Sun distance, or about 11.2 light-years. Shortly after Bessel's discovery, the astronomer Thomas Henderson detected a parallax in the closest star, alpha-Centauri, of 0.76 seconds of arc, placing it 4.3 light-years away.

Thus, by the mid-nineteenth century, the distance to a few of the closest stars was finally known. By measuring stellar parallax, our view of the size of the universe increased by nearly a millionfold, from the mere seventy-five-million-mile universe of Ptolemy to the sixty-six-million-million miles to the star 61-Cygni. But only the distances to the closest stars could be determined by parallax. The vast majority of stars are so far away that their parallaxes are too small to measure even with the best telescopes of today.

This first step in the measurement of distances to the stars was in some ways very modest; even today it takes us out to distances of about three hundred light-years, a tiny fraction of the size of our galaxy. However, the old adage that even the most distant journey begins with a small step was never more apt, for without this first step, we would be unable to begin the journey to distant galaxies.

It is sobering to note that of the one hundred billion or so stars in our galaxy, only about a thousand are close enough to measure a parallax. It's not impossible to imagine that we could live in a part of a galaxy with *no* stars close enough to measure a parallax. In that case we would never have been able to determine the distance to the few nearby stars necessary to make that first small step in the third dimension, and those little twinkling points of light in the sky would forever remain a mystery.

LINES IN THE NIGHT

The determination of distances to stars illustrates that there is no single way that science proceeds. The first step in the process of the exploration of the third dimension was the establishment of a conceptual framework to gauge distances in the solar system—in other words, constructing a model. After Newton's synthesis of gravity, Copernicus's model was clearly the one to use, and the arrangement and distances to objects in the solar system were finally understood. Of course, further refinements in the determination of the distances were needed, but the basic idea was in place.

The next step to the stars was the search for stellar parallax. This was an example of a long and arduous search for an effect predicted by a well-accepted theory. Everyone was looking for it, and it was only a matter of time (nearly two centuries in this case) before instruments were refined enough to detect it. By the end of the nineteenth century, after the discovery of parallax, it was realized that although there were millions of stars to be seen with telescopes, only a handful were close enough for detection of a parallax. But no one had a clue how to judge the distance to stars too far away to have a measurable parallax. This step involved a different aspect of scientific progress—the slow, patient accumulation of facts in the hope that something useful would result. This involved studying tens of thousands of stars searching for some way to extend the distance scale beyond the nearby stars. The search required patient work and the introduction of two new tools into astronomy: photography and **spectroscopy**.

Just as in the dark all cats are gray, in the dark night sky all stars are more or less white. On a clear night, however, slight differences in color among the stars can be noticed. For instance, if you look carefully on a good night you will notice that Betelgeuse, a bright star in the winter

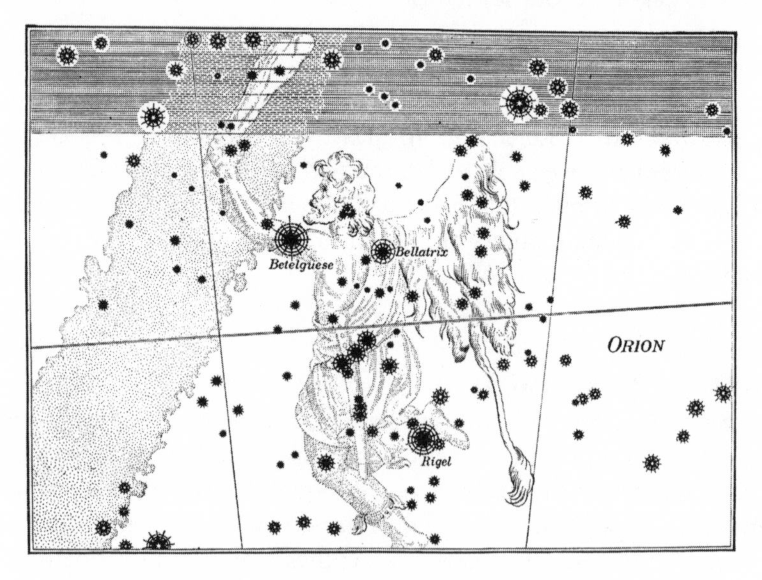

The constellation Orion: The star Betelgeuse is reddish, and the star Rigel is blue-white.

constellation Orion, appears reddish compared with a nearby star in the same constellation, the blue-white star Rigel. Gathering more light by using a telescope or binoculars accentuates the apparent color differences, but color is often a subjective attribute. This subjectivity was removed by the introduction of spectroscopy in the second half of the nineteenth century.

✦ ✦ ✦

The study of the spectrum of the light emitted by an object is an important tool in many fields of science: chemistry, physics, and astronomy to name a few. One of the natural phenomena that fascinated Newton was that sunlight is dispersed into different colors when passed through a prism. In the beginning of the nineteenth century, several people, most notably the German optician Joseph von Fraunhofer, discovered that the spectrum of light from the Sun wasn't continuous, but on magnification hundreds of fine, dark lines could be seen.

Along with Fraunhofer, Wilhelm Bunsen (of burner fame) and Gustav Kirchoff established the science of spectroscopy. By the middle of the nineteenth century, it was known that

1. there are more than five hundred dark lines in the spectrum of the Sun (by the end of the nineteenth century more than twenty thousand lines were identified);
2. the light from the Moon and the planets has the same pattern of dark lines as the Sun;
3. a glowing solid body produces a continuous spectrum without the dark lines seen in the solar spectrum;

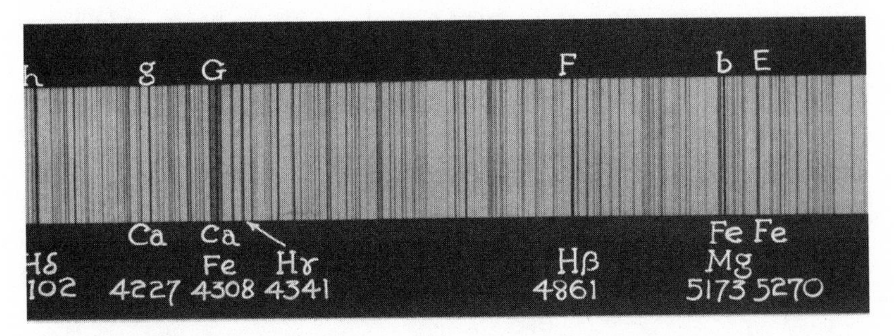

A part of the solar spectrum showing the dark lines. By measuring the position of the dark lines, one can determine the chemical composition of the sun.

4. a glowing gas emits light with a spectrum consisting of a series of thin bright lines;

5. if a continuous spectrum of light from a glowing solid is passed through a gas, dark lines appear in the spectrum;

6. the positions of the lines in the spectrum depend on the gas, so they can act as fingerprints that allows identification of the chemical composition of the material emitting or absorbing the light.

By 1859 Kirchoff knew enough about the spectra of gases from laboratory studies to identify the chemical elements in the Sun responsible for the dark lines in the solar spectrum. Thus, on the basis of experiments done on Earth, he could discern that the Sun is not made of any heavenly substance like quintessence but of everyday earthly elements. The accomplishment is remarkable in many ways. About twenty-five years previously the French philosopher Auguste Comte, founder of Positivism, had confidently stated about the Sun and the stars that "we can never by any means investigate their chemical composition...."[5]

I often wonder why history doesn't take more notice of Kirchoff's accomplishment. The idea that we learned what the Sun and the stars are made of would have astonished the ancients—it still astonishes me. Some philosophers and historians are so alienated from science that the significance of the discovery is hardly mentioned. This was made painfully clear to me one spring day in 1989, when, during a banquet at a physics conference in Rome, I found myself sitting next to a physicist's spouse who happened to be a historian at the University of Rome. Although astronomy is a highly specialized profession, I am always amazed by the degree of specialization in other fields. She was an expert on European history of the year 1859 (presumably the university has one hundred nineteenth-century European historians). In a clumsy attempt at polite dinner conversation, I asked why she happened to concentrate on that year. With a "surely you must know" tone, she replied that it was a very significant year because of the development of a remarkable idea. I made the mistake of asking if she was referring to Kirchoff's discovery of the chemical composition of the Sun. She stared at me so long, with such a curious expression on her face, that I thought surely I must have linguini stuck to my chin. But no, she was simply amazed by the naïveté of my question. Finally, she informed me that the

[5] It is hard to judge what is more unreliable—a philosopher's list of what can never be discovered or a scientist's list of what is certain to be discovered.

significant event of the year 1859 was the publication of *A Critique of Political Economy*, by Karl Marx. I further compounded my errors by asking how a mere economic theory could be compared to the discovery of the composition of the stars. I suppose that a biologist might ask why she considered Marx's book more important than another book published in 1859, *On the Origin of Species*, by Charles Darwin. After another long stare, with a sigh of exasperation she turned to the person sitting on her other side, presumably searching for more enlightened conversation. I am embarrassed to admit that in the intervening years I still haven't understood why the development of a (since discredited) economic theory is of more lasting importance than learning the stuff of which the stars are made. Perhaps one day I will.

I also wonder why the significance of scientific discoveries is so often dismissed by historians in favor of political, military, or economic developments. As noted by Arthur Koestler, in Somervell's abridged version (if more than six hundred pages can be considered abridged) of Toynbee's *A Study of History*, the names of Copernicus, Galileo, Descartes, and Newton do not appear.

It wasn't long before astronomers were examining the spectra of the light from stars. In the 1860s the Italian Jesuit Pietro Angelo Secchi noticed that stars of different colors had different spectra, and published a list of stars in which he classified them in four groups according to their spectra. Stars in the first class were bluish, stars in the second class were yellow, and stars in the final two classes were both red but with very different spectra. By 1890 a dedicated team at the Harvard College Observatory led by Annie Jump Cannon and Antonia Maury published the Henry Draper catalog of the spectra of 10,000 stars, and by 1924 the catalog was extended to 225,000 stars. In the first Draper catalog the stars were divided into seven types denoted by the letters O, B, A, F, G, K, and M, with O and B corresponding to Secchi's first type; A, F, and G to his second; K to his third; and M to his fourth.[6]

This was truly a tremendous undertaking involving careful, painstaking research without any obvious immediate goal other than

[6] Of course, the number of categories grew as more stars were classified. Now there are 13 main classes denoted by the letters Q, P, W, O, B, A, F, G, K, M, S, R, and N, and each class is subdivided into ten groups designated by the digits 0 to 9 placed after the letter. The Sun is a G2 star.

classification. One has to admire the dedication of the people working on the catalog because there was no obvious immediate scientific payoff. They had no way of knowing whether spectral classification would lead to a deeper understanding of the nature of stars, for until the theory of stellar evolution was developed, no one understood the significance of the different spectra. It was similar to zoology before the development of the theory of evolution. Without the organizing principle of natural selection and evolution, the division of animals into species, genera, phyla, and so on is classification without real understanding. Without the organizing principle of stellar evolution, sorting the stars by spectra could be said to be classification without real understanding.

But the classification paid off when it was discovered that there is a correlation between the spectral class and the intrinsic brightness, or **luminosity**, of stars. Although it was a remarkable insight at the time, that one should look for a trend of spectral class with brightness seems logical in hindsight. After all, there are basically only two things you can notice about a star—how bright it is and its color—so to try to find a correlation of brightness and color makes sense. The correlation is usually demonstrated by observing a large number of stars, and then making a graph where each star is represented by a point with brightness on one axis and color on the other.

Well, let's try it. Plot the brightness of the star on one axis and the color of a star on the other axis, and lo and behold...a mess. Of course it's a mess, because we plotted the wrong thing. The apparent brightness of the star depends on two quantities—the intrinsic luminosity of the star and its distance from us. A star will appear brighter if it is closer to us than if it is farther away. So what we want to plot is the intrinsic luminosity of the star on one axis and the color on the other, because it is the intrinsic luminosity that is a property of the star and doesn't depend on where it is located.

But we don't measure directly the intrinsic luminosity of a star. When we view a bright star in the night sky, we can't tell immediately whether it is bright because it is nearby or because it is far away but has an intrinsically large luminosity. If we somehow know the distance to the star by some other means, however, we can tell whether the star is intrinsically bright or dim. Of course, we do know the distance to nearby stars by parallax, so for them we can determine their intrinsic luminosity and plot it against the color of the star.

This was done in 1914 by the American astronomer Henry Norris Russell (and somewhat earlier on a more modest scale by the Dutch

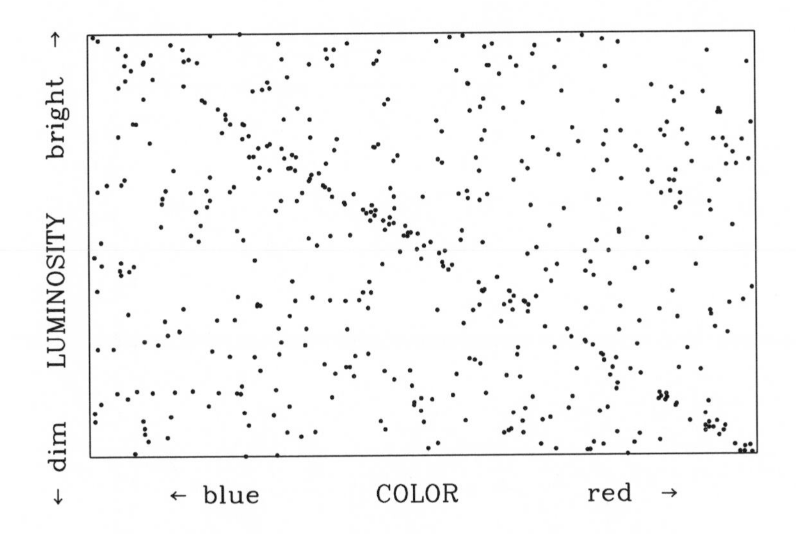

A graph of apparent *brightness and color for five hundred of the closest stars. There is no obvious correlation because the apparent brightness of a star depends on both its intrinsic brightness and its distance. To make sense of this jumble of information, one must know the intrinsic luminosity of the star, and not just its apparent brightness. This requires knowledge of the distance to the star.*

astronomer Ejnar Hertzsprung) for the three hundred stars whose distances were known at the time by their parallax. From knowledge of their apparent brightness and distance, Russell deduced their intrinsic luminosity and plotted it along with their spectral type, known from the Draper catalog. The result from Russell's 1914 paper is shown in the illustration on p. 166.

An obvious trend can be seen in Russell's graph. Blue or white stars are bright (although there are some dim white stars that Hertzsprung called white-dwarf stars), and most red stars are dim (although there are some red luminous stars that Hertzsprung called red giants). The vast majority of the stars lie between the two lines shown in the illustration, so for most stars there is a known relationship between color and intrinsic luminosity. Therefore by measuring the color (or more precisely the spectral class) of a star, you can determine its intrinsic luminosity.

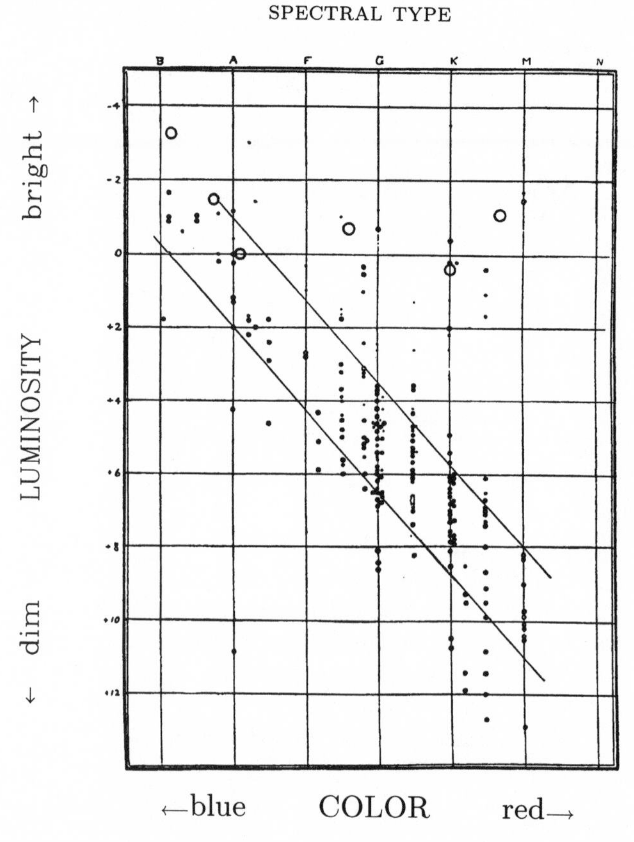

The Hertzsprung-Russell diagram: This original illustration from Russell's 1914 paper is the graph of the intrinsic luminosity of stars versus their color (or spectral type). The large open circles are for stars of uncertain parallax, hence uncertain intrinsic luminosity. The intrinsic luminosity at the top of the diagram is 7,500 times that of the sun; the lowest point on the scale is 1/5,000 of the solar luminosity.

Once the intrinsic luminosity of the star is known, it is easy to calculate how far away the star must be to have its apparent brightness. If we know that two stars have the same intrinsic luminosity and one appears four times brighter, then by the inverse-square law the fainter one must be twice as far away as the brighter one. Thus, by measuring their color and apparent brightness, we can determine the distance to stars that are too far away to determine their distance by parallax.

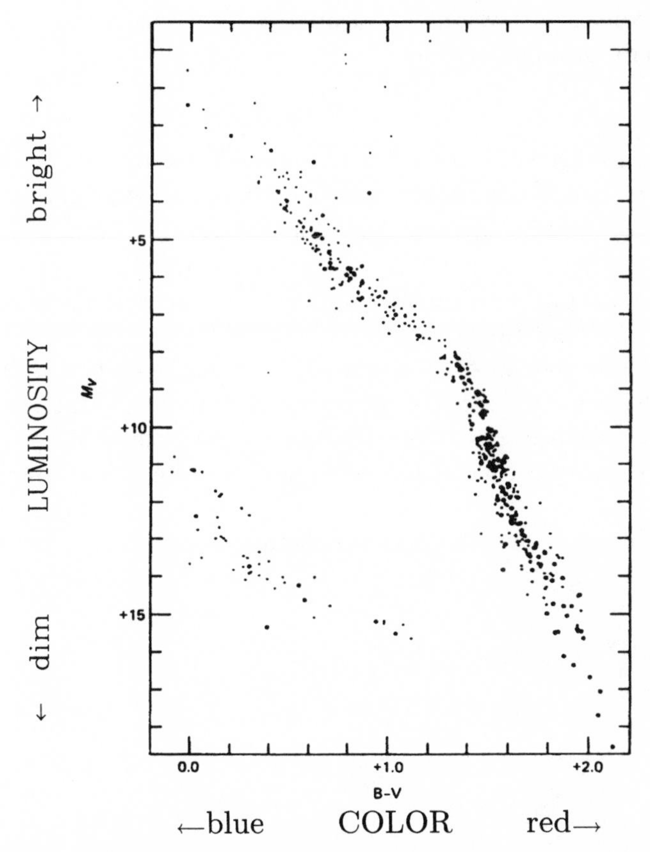

A modern Hertzsprung-Russell diagram for five hundred of the closest stars. The color of the star is defined as the difference in the brightness when the light from the star is passed through two different filters; one filter transmits only light in the green region of the spectrum, and the other filter passes only light in the violet region of the spectrum. Most of the stars lie on a curve running from upper left to lower right.

This step on the distance ladder can take us nearly across our galaxy. Although a giant step, it could be taken only because Hertzsprung and Russell were able to find the relationship between intrinsic luminosity and spectral type, for which they needed to know the distance to nearby stars by their parallax. The distance to stars by parallax depends on the distance from Earth to the Sun, which Aristarchus first determined in terms of the Earth–Moon distance, which he knew in terms of

Earth's radius, which was first measured by Eratosthenes with the aid of a hole in the ground in Aswan.

✦ ✦ ✦

Twenty-five years ago I thought Elnora's question about the distance to the Moon was one of the funniest things I had ever heard. But a lot of things amusing to a seventeen-year-old seem less humorous as the years pass. After many years struggling to understand the cosmological distance scale, I now realize the question should have been treated with respect rather than ridicule.

It is now too late to apologize to Elnora for laughing at her question. But when confronted in the course of my work with uncertainty about the distance scale, I often reflect on the irony that I spend so much time worrying about a subject I once thought was so straightforward.

Never again will I laugh at "simple" questions.

SEVEN

Islands in the Sky

It seems that every week or so one reads of some new, remarkable scientific discovery. For example, seventeen times in the past decade I have read the exciting news that someone has found the first evidence for the existence of black holes. More claims have been made for the first discovery of a black hole than alleged first discoveries of the Americas. It is difficult for scientists reading newspaper and magazine reports about their own field to predict which discoveries are significant; it is probably impossible for the nonexpert to separate the meaningful discoveries from the more mundane day-to-day progression of science.

Although the significance of a discovery is often impossible to discern immediately, it is a simpler matter to tell how hard the discovery was to make. The most difficult thing to discover is something you are not looking for, and the easiest thing to discover is something you *are* looking for. "Seek, and ye shall find," may have been intended as solace for the spiritually deprived, but it must be interpreted as a serious warning by the scientist. For it is human nature that if you look hard enough for something, you want very much to find it, and you work at the very edge of the limit of your instruments, you have to try very hard *not* to find it—whether or not it is there. Many scientists have seen things that they were looking for but that weren't really there.

Indeed, some of the greatest scientists have erroneously confirmed the existence of nonexistent phenomena. Among the greats who have erred in this way were Galen and Leonardo.

During the second century of the Christian Era, about the time the Greek astronomer Ptolemy was watching the skies in Alexandria, Clarissimus Galen, a Greek physician living in Rome, was studying human anatomy. A patient and accurate observer of nature, he conducted anatomical investigations that were unrivaled in antiquity and hardly surpassed as late as the seventeenth century. For thirteen hundred years, just as Ptolemy was considered the authority on all things astronomical, Galen was regarded as the authority on matters anatomical. Just as in the case of Ptolemy, Galen's reputation was well deserved. He was responsible for a vast number of discoveries about the human body: he demonstrated that the arteries contain blood and not air as popularly believed at the time, he made noteworthy contributions to neurology, and in an age when most physicians jealously guarded their knowledge lest another physician discover it and steal their patients, he wrote hundreds of treatises on medicine, along with many on philosophy, comedy, and logic.

In the curious Roman world where the slaughter of gladiators was considered wholesome family entertainment, the public recoiled with horror at the thought of medical examination of the losers, so most of Galen's information was the result of dissection of apes and other animals. The bulk of his anatomical observations are true to nature, but some are curious exceptions. Galen believed, as did many physicians of his time, that women had two uterine cavities, one on the right for the male fetus and one on the left for the female fetus. But what concerns us here was his "model" of the human heart.

In *De Usa Partium*, Galen described the anatomy of the human heart and the circulatory system. For well over a thousand years this was considered the final word on the heart and circulation of the blood, and not until William Harvey's great work in 1628 was it supplanted. But in his description of the heart, Galen erroneously reported that the septum was permeated by a multitude of barely perceptible formina, through which some of the blood exuded from the left ventricle to the right ventricle.

The study of human anatomy in the Western world did not advance much further for a thousand years after Galen, in part because in the equally curious Christian world, dissection for the purpose of the

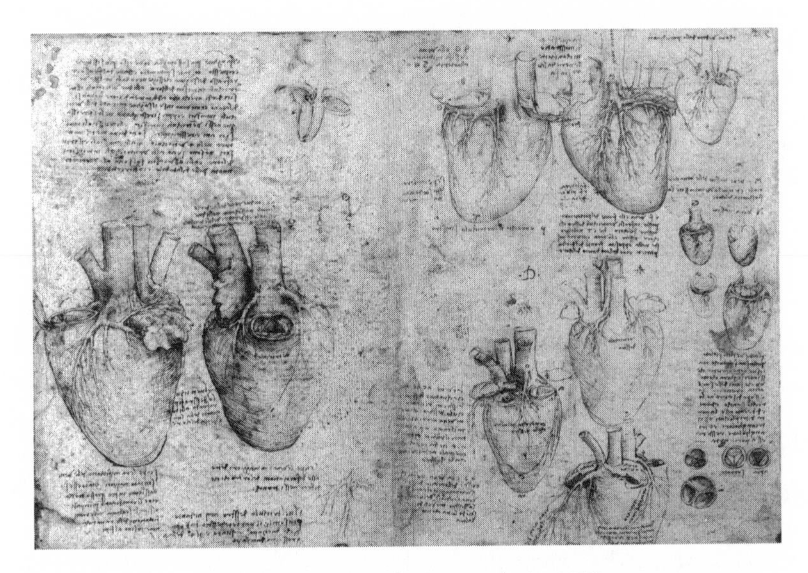

The heart as seen by Leonardo in 1513.

acquisition of scientific knowledge was forbidden on the grounds that it was impious to mutilate an image of God. (Yet there was little hesitation about crushing the bones or burning the flesh of thousands upon thousands of images of God thought to be witches or heretics.)

Interest in human anatomy was rekindled in the Renaissance. The action and purpose of the human heart could not escape the relentless curiosity of Leonardo da Vinci, who was one of the first of the modern dissectors until Giovanni di Medici, Pope Leo X, banned him from the Roman hospital Ospedale di Santo Spirito, calling him "a heretic and cynical dissector of cadavers." So, like Galen, many of Leonardo's dissections were on animals. Leonardo was well on his way to discovering the principle of circulation of the blood, but as in the case of so many of his other investigations, as well as art, he seemed pathologically unable to bring it to a final conclusion.[1] Nevertheless, his study of the heart was without parallel among his contemporaries. His pen and brown ink drawings of the heart show an incredible attention to detail, for the most part rivaling anything to be found in *Gray's Anatomy,* and in its own way possessing the same sublime beauty as the smile of the Mona Lisa.

[1] On another occasion Leo X said of Leonardo, "This man will never finish anything, for he starts by thinking about the end before the work is begun."

It is well known that Leonardo was fiercely independent of authority in scientific investigations. In words that might have come from Galileo a hundred years later, he expressed his feelings in his famous notebooks:

> I do not understand how to quote from learned authority, but it is a much greater and more estimable matter to rely on experience. They scorn me who am a discoverer; yet how much more do they deserve censure who have never found out anything, but only recite and blazon forth other people's works. Those who study only old authors and not the works of nature are stepsons, not sons of nature, who is mother of all good authors.

Yet, despite his declaration that he did not rely on the old authors, when Leonardo depicted the heart, he included minute pores in the interventricular septum that were placed there by Galen, not by nature. But Leonardo drew the partitions not because he "trusted" Galen as the supreme authority, but because he thought he actually saw them. No less an acute observer of nature than Leonardo da Vinci at times saw what he expected to see, not what was really there.

Leonardo was no blind watcher of the heart, but he was working at the limits of the resolution of his instrument, the human eye. When he held a heart in his hand, he expected to find holes that were just barely large enough to see, and he found what he expected.

When astronomers place their eye to the eyepiece of a telescope (or in modern times gaze at a video monitor), they are often looking for objects just at the limit of their instruments. That is why forefront areas of research in astronomy often seem like a morass of contradictory observations, confusing statements, and wild speculation. But in the progress of science, a truer picture of nature inevitably will emerge to replace our most cherished illusions. There is probably no better example in astronomy of the triumph of truth over illusion than the issue of the nature of nebulous regions of the sky.

HEAVENLY CLOUDS

In addition to the sharp images of stars and planets, the naked eye can discern regions of the sky that are nebulous, or cloudy. These are often referred to by the generic name *nebula* (plural form *nebulae*, with the last syllable pronounced "lee"), which comes from the Latin word for "cloudy." The most prominent nebulous region of the sky is the Milky Way, a white, diffuse region of the night sky that can be seen on most clear nights of the year stretching across the entire horizon.

The Milky Way was known to the ancients, but they showed little interest in it. Aristotle believed that it was sublunar, that is, located somewhere between us and the Moon, probably some sort of atmospheric phenomenon. Ptolemy treated it, but not at great length, and Copernicus barely mentioned it.

The true nature of the Milky Way was not glimpsed until Galileo turned his optic tube in its direction and saw that it appeared cloudy because there were stars "so numerous as almost to surpass belief." As Galileo wrote in *The Starry Messenger*,

> [The Milky Way] is in fact nothing but a congeries of innumerable stars grouped together in clusters. Upon whatever part of it the telescope is directed, a vast crowd of stars is immediately presented to view. Many of them are rather large and quite bright, while the number of the smaller ones is quite beyond comprehension.

But the Milky Way was not the only nebulous region in the sky known to astronomers before the invention of the telescope. Among the handful of visible nebulae was one in the constellation Andromeda, to the unaided eye dimmer than the Milky Way, with an apparent diameter about as large as the Moon. We now know that this is the Andromeda galaxy, the closest large galaxy to us, about two million light-years distant. Simon Marius, one of Galileo's contemporaries (and to Galileo every contemporary was a rival), claimed to have resolved the Milky Way before Galileo, and went on to claim that he was able to see many distinct stars in the Andromeda nebula.

Ah, but poor Marius, working at the limit of his instrument, found stars in Andromeda placed there by his imagination, not by nature. There are indeed individual stars in the Andromeda nebula, but they are far too distant to be seen by the eye, or even with telescopes before the twentieth century.

Scanning the heavens with their new tool, astronomers discovered many new nebulae in the early era of the telescope. Galileo believed that all nebulae were groups of stars that only appeared nebulous because they were either too far away or too close together for him to resolve the individual constituent stars. The question of whether *all* nebulae were simply collections of stars that could be resolved with sufficiently powerful telescopes was debated for centuries.

Just as for the stars, when faced with any new phenomenon in nature the first thing to do is to collect as much information about as many examples as possible and try to classify the different kinds. In 1733 the English astronomer William Derham listed twenty-one nebulae (only seven of which are actually nebulae). Another list of thirty nebulae was put together by the Swiss astronomer Jean-Phillipe Loys de Chèseaux in 1746. The number of known nebulae slowly began to grow as telescopes improved in the eighteenth century.

The most complete early listing of nebulae was compiled in the 1780s by Charles Messier. The Messier catalog consisted of 103 entries, 42 of which he discovered himself. The Messier catalog proved very important to the development of our understanding of nebulae, despite the fact that Messier himself was not particularly interested in the subject. His main interest was hunting for comets, and his motivation for assembling the list of nebulae was to prevent people from mistaking them as comets.

The nature of nebulae remained a fundamental question in astronomy until 1924. The study of nebulae attracted some of the greatest astronomers from the time of Galileo until today, and confounded most of them.

✦ ✦ ✦

Part of the confusion in the development of our understanding the nature of nebulae was the expectation that every nebulous object seen in the sky was a different manifestation of a single phenomenon. One of the usual hallmarks of a successful scientific theory is that a single assumption or explanation accounts for a variety of seemingly different phenomena. Everyone is familiar with the story of the blind men and the elephant. In most instances, a great scientific discovery is akin to the realization that the men are holding different appendages of the same animal. However, this was not the case for the nebulae; the variety of things called nebulae actually were different animals, not different appendages of the same animal. It was as if astronomers were blind men holding on to parts of different animals, thinking it was a single animal, and trying to imagine what it looked like.

To appreciate better the unfolding of the saga of the nature of the nebulae, it is helpful to know the final answer. Now we know that catalogs that once tossed together everything into a single object known as a nebula contained many different creatures, as distinct as these:

1. *Diffuse nebulae*, often just referred to as nebulae, which consist of regions of dust or gas in the interstellar medium illuminated by the light emitted by a nearby bright star (or, more typically, several stars).

Examples of Nebulae Listed in the Messier Catalogue

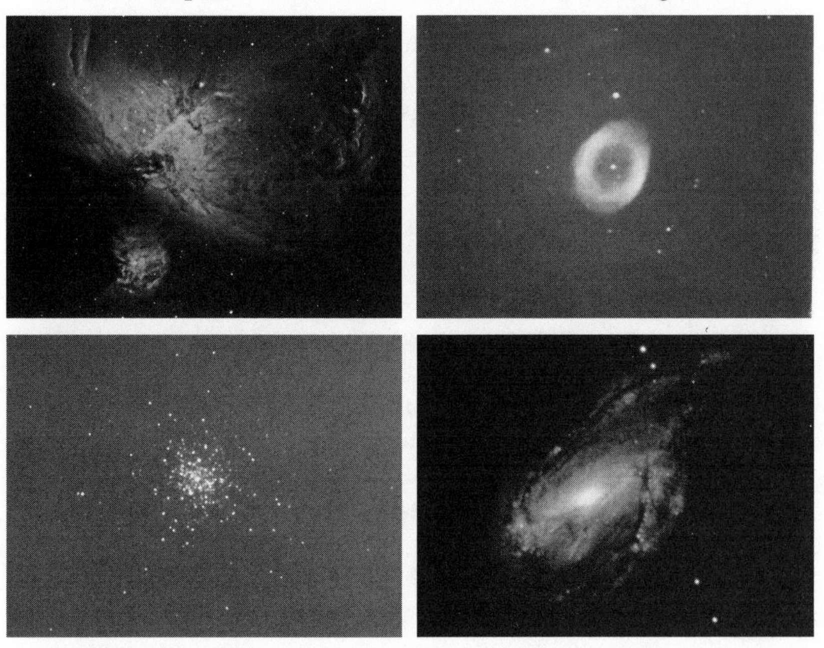

The upper left is a region of the Orion nebula, the forty-second entry in Messier's catalog. Gas in the cloud is glowing by re-emitting the energy received from several nearby hot, young stars. It is about fifteen hundred light-years away. Shown in the upper right is a planetary nebula known as the ring nebula, the fifty-seventh entry in Messier's catalog. It has a single bright central star, illuminating gas that was ejected by the star about twenty thousand years ago. It is about twenty-three hundred light-years away. At the lower left is the third entry in Messier's catalog, a globular cluster consisting of about two hundred thousand stars about forty-three thousand light-years away. Finally, in the lower right is a spiral galaxy, Messier's sixty-fourth entry in his catalog. It consists of about one million-million stars at a distance of about thirteen million light-years.

2. *Planetary nebulae,* which are relatively thin shells of gas surrounding a single star. The gas was blown off from the star and is expanding away from it at a high velocity, illuminated by the light from the star.

3. *Star clusters,* made up many individual stars held together by the force of gravity. Small, loose groups of stars barely large enough to hold themselves together by gravitation are known as open clusters. Larger bound groups of stars, containing as many as a million stars in a tightly packed spherical region, are known as globular clusters.[2]

4. *Galaxies* external to our own Milky Way galaxy were first identified by their nebulous, nonstarry appearance. Galaxies are large assemblies containing as few as one hundred million stars, or as many as ten thousand billion stars. Galaxies may also contain all the above types of nebulae.

Of the original 103 entries in Messier's catalog, 7 are diffuse nebulae, 4 are planetary nebulae, 28 are open clusters, 29 are globular clusters, 34 are galaxies, and 1 is a double star that didn't really belong in the catalog. Of the 103 objects, 69 are galactic, and 34 are extragalactic, or outside of our Milky Way.

Some examples of the various objects known in the eighteenth and nineteenth centuries by the common name *nebula* are shown in the illustration. Objects we now know to be as small as a single star, or containing as many as ten trillion stars (and just about everything in between) were lumped together as nebulae. The Messier catalog contains things now known to be as nearby as five hundred light-years, or as distant as one hundred million light-years.

Today we know the difference between a galaxy and a planetary nebula, but it required a couple of hundred years to straighten it out. Examining the sample of very different objects from the Messier catalog, and imagining astronomers looking at their fuzzy images through eighteenth- or nineteenth-century telescopes without the benefit of photography, one might easily imagine that it was not so clear at all that they represented very different types of animals.

Of course, at the time Messier didn't know what he was seeing; he knew only that they weren't comets. He would have been truly astonished to learn that he was viewing objects outside of our galaxy, as distant

[2] Groups of stars not bound together by gravity are called "associations."

as one hundred million light-years. The story of how we finally discovered that some of the nebulae were galaxies, known in the early days as "island universes," is another chapter in our struggle to plumb the depths of the night sky.

✦ ✦ ✦

That the Milky Way was the first nebula to be widely studied and discussed is not surprising. Although the development of our picture of the structure of the Milky Way is the subject for another book, there are a couple of very interesting things about the mapping of our galaxy. The fact that our galaxy is shaped like a flattened disk emerged rather early in the game, and after the work of Thomas Wright in 1750, Immanuel Kant in 1755, and Johann Lambert in 1761,[3] most astronomers were prepared to accept that our galaxy, which was thought by most people to be the same as the entire universe, is not spherical after all.

However easy it was for astronomers to alter their prejudice as to the shape of our galaxy, they were much slower to accept the fact that our solar system is not the center of the galaxy. Not until 1918 was this demonstrated beyond doubt; until that time it was generally assumed that our solar system is at the center of the universe simply because of the absence of any evidence to the contrary.

The question of whether all the other nebulae were just stars was a tougher nut to crack. An example of the difficulty of discerning the nature of the nebulae from the fragmentary and contradictory pieces of information available to early astronomers is illustrated in the life and work of the Herschels, William and John.

RESOLVING THE ISSUE

William Herschel was born in in Hanover, Germany, in 1738. In 1753 he joined his father as a musician in the Hanoverian guard. He first arrived in England when his regiment visited there in 1757. Later that year he resigned his commission and settled in England to pursue a career as a musician, music teacher, and composer. Although not as gifted as George Frideric Handel, another German expatriate living in England, Herschel was noted for a number of musical accomplishments: among

[3] An amusing fact is that all three devoted almost as much space to the discussion of intelligent extraterrestrial life as to the discussion of the shape of the galaxy. Lambert went so far as to propose that comets were inhabited.

his major works were twenty-four symphonies, seven violin concertos, and two organ concertos. His musical output was impressive, but his astronomical output was truly prodigious.

Herschel could have had a happy and successful life as a musician, but like Tycho and so many before him, he was captivated by the stars. The first entry into an astronomical diary by the musician and (at the time amateur) astronomer was in 1766. A scant ten years later he completed the building of one of the finest telescopes of the time, a twelve-inch aperture instrument, followed by progressively larger ones, culminating with a forty-eight-inch telescope completed in 1788, mostly underwritten by funds from the Crown.

Herschel is popularly known for the discovery of the planet Uranus, the first planet discovered with the aid of a telescope. Actually, at its brightest Uranus is just barely discernible by eye, but it had escaped detection because there are thousands of other objects of similar apparent brightness. In fact, on at least two dozen previous instances astronomers had recorded the position of Uranus, thinking it was a dim star. But Herschel was simply a more careful watcher of the sky.

Perhaps his early musical training had engendered in him an ability to "sight-read" the sky, for in the style of Tycho Brahe, Herschel set

Sir William Herschel (1738–1822) (left), and his son, Sir John Herschel (1792–1871) (right).

about to survey the sky, recording the positions of ever dimmer objects. On March 13, 1781, while sight-reading a familiar region of the sky with his telescope, he noticed a dim object that had not been there before. Such an inconspicuous object might have escaped the attention of a less acute observer, but it caught Herschel's eye. Later, when writing about his ability to see, he explained that it was no accident, but the result of practice:

> Seeing is in some respects an art that must be learnt. To make a person see with such power [as I am able] is nearly the same as if I were asked to make him play one of Handel's fugues upon the organ. Many a night have I been to see, and it would be strange if one did not acquire a certain dexterity by such constant practice.

Herschel was overjoyed to think that he had discovered something—not a planet, but a comet! In fact, the first paper by Herschel announcing the discovery of Uranus was titled "Account of a Comet." In the paper Herschel reported that during the first six weeks of viewing, the object had doubled in size, exactly as expected for an approaching comet. We now know, however, that during that period in 1781 the apparent size of Uranus should have been slightly decreasing.[4]

Herschel soon came to realize his error and appreciated that he had discovered a planet. Herschel wanted to name the planet "Georgium Sidus" (George's star), after his fellow Hanoverian, George III, king of Britain. This obvious attempt to curry favor paid off when King George offered to fund Herschel's research and to underwrite the construction of an even larger telescope.[5] Perhaps learning that a new planet had been named in his honor was better news to the king than that from the Americas, where in the year of Herschel's discovery Cornwallis surrendered at Yorktown. Of course astronomers on the Continent would not countenance a planet named for a British monarch, and the planet eventually came to be known as Uranus, named for Urania, the muse of astronomy.

[4] In planetary astronomy Herschel also discovered two satellites of Uranus and two new moons of Saturn. His reported detection of an additional four moons of Uranus was later shown to be spurious. Herschel also held the curious belief that the Sun is habitable. As we know, even the greatest astronomers, with the ability to see what others can't, occasionally are blind watchers of the sky.

[5] As with Fredrick II of Denmark and Rudolph II of Bohemia, astronomy always seemed to fare well under rulers considered mentally unstable.

As part of his sky survey, Herschel turned his large telescopes toward the nebulae and read their locations on the celestial sphere. In 1786 he published a catalog of sky positions of nebulae, as several predecessors had done. But just six years after Messier's catalog of 103, Herschel's catalog contained a thousand objects. Just three years after that he added another thousand to the list, followed by a final list with an additional five hundred in 1802.[6] Truly this was a testament to Herschel's patience and ability, as well as the power of his instruments.

One question Herschel tried to answer was whether nebulae could be resolved into stars, or whether they are made out of some non-starry, diffuse material. Now we know that this question should be prefaced with the stipulation of exactly what kind of nebulae. But guided by the fickle spirit of simplicity, Herschel thought all nebulae were the same animal.

In 1784 Herschel published his first important paper about the nebulae. With his new nineteen-inch telescope he examined many of the objects identified by Messier. Since most of the objects in Messier's catalog were either open clusters or globular clusters, they are resolvable into stars by instruments like Herschel's. He identified stars in 29 of the 103 Messier objects, and reported that he expected to be able to find stars in 9 others.[7] The tone of the 1784 paper is that all nebulae are resolvable with a sufficiently large telescope, and he was confident that it would be only a matter of time and effort until he did just that.

But his tune quickly changed. In a paper the following year, Hershel proposed that his catalog of nebulae could contain different species of animals. He suggested classifying them into five types, and coined the terms *planetary nebula* and *globular cluster* in the process. Not all his classifications survive, but it was certainly a step in the right direction. In the paper he also cautiously backed away from his earlier expectation that eventually all nebulae would be resolved, and suggested that planetary nebulae at least were surrounded by true nebulosity that would never be resolved into stars.[8]

[6] The vast majority of the objects discovered by Herschel were external galaxies.

[7] We now know that of the nine objects, only two would have been resolvable by Herschel's telescopes.

[8] Curiously, despite his guess that the nebulosity surrounding planetary nebulae could never be resolved into stars, in the same paper he incorrectly claimed to have resolved into stars the archetypal planetary nebula, the ring nebula (shown earlier in the illustration of examples from Messier's catalog).

As Herschel trained progressively larger telescopes skyward he discovered more and more nebulae, and whereas the objects in earlier catalogs like Messier's were predominantly galactic, the bulk of nebulae discovered with larger telescopes—such as Herschel's giant telescope containing a mirror four feet in diameter finished in 1788—were extragalactic. It would be more than a century before individual stars in these external galaxies would be identified, and of course Herschel had no idea that they were extragalactic.

By 1791 Herschel had come full circle on the question of whether all nebulae are made of pointlike stars, and in a paper, "On Nebulous Stars, Properly So Called," he said as much. In November 1790 he had observed a planetary nebula with his recently completed forty-eight-inch telescope, and that seemed to leave no doubt in his mind that the diffuse glow around the central star would never be resolved into stars.

In his final major paper on the subject, written in 1811 when he was seventy-three years of age, he freely admitted his change of opinion:

> We may also have surmised nebulae to be no[ne] other than clusters of stars disguised by their very great distance, but a longer experience and better acquaintance with the nature of nebulae, will not allow a general admission of such a principle, although undoubtedly a cluster of stars may assume a nebulous appearance when it is too far remote for us to discern the stars of which it is composed.

By the time of his death in 1822, Sir William Herschel had bequeathed to astronomers catalogs listing the positions of more than twenty-five hundred nebulae, built the greatest telescopes of his age, made the first attempts to classify nebulae, discovered the planet Uranus, been one of the founders of stellar astronomy, confirmed the disk-like structure of the Milky Way, and changed our view of the nebulae from annoying little objects interfering with the search for comets to important astronomical subjects worthy of study. He also left to astronomy a son, John Herschel, later Sir John Herschel, who extended from an observatory in South Africa his father's sweeps of the sky to regions not visible from England, adding 1,269 new nebulae.

Although William Herschel left the study of nebulae in nearly as confused a state as he found it, there were occasional blind spots in his sweeps of the sky, and there were others that contributed to the development of our understanding of the nebulae, he was a brilliant astronomer who saw things others did not. Occasionally he saw things placed in his eyepiece by his imagination and not by nature, but he had the strength of character to change his opinion and the integrity to admit when he was wrong. The final two attributes alone make him a most unusual person.

The study of objects outside our galaxy requires large telescopes, and the construction of large telescopes requires a large amount of capital. After the death of the elder Herschel, and the accompanying disappearance of Crown funds, the construction of large telescopes passed to

The Growth of Telescopes

The top pictures are telescopes constructed by William Herschel, a nineteen-inch completed in 1783 (left), and a forty-eight-inch completed in 1788 (right). On the bottom are telescopes of Lord Rosse, a thirty-six-inch finished in 1840 (left) and the Leviathan of Parsonstown, completed in 1845 (right). The Leviathan had a four-ton copper and tin mirror. Because of its enormous weight, it was mounted between sixty-foot tall brick walls and could only be tilted up and down, not rotated from side to side.

the third earl of Rosse, William Parsons, an Irish nobleman. Lord Rosse had the personal resources necessary for the construction of large telescopes, including a thirty-six-inch instrument in 1840, and culminating with an enormous seventy-two-inch monster known as the "Leviathan of Parsonstown," completed in 1845. The Leviathan was an ideal instrument for the study of the nebulae.

In his investigations of nebulae with the Leviathan, Rosse took a giant step forward, as well as a giant step backward. The backward step was the mistaken impression that he could resolve all the nebulae. Rosse had "little if any doubt" that he could resolve the unresolvable Orion nebula, and claimed to resolve into stars the fifty brightest nebulae in the Herschels' catalog. This led John Herschel to state that all nebulae are in principle resolvable. Another astronomer, J. P. Nichol working in Glasgow, put it in even stronger terms when he wrote, "Every shred of that evidence which induced us to accept as a reality, accumulations in the heavens of matter not stellar, is forever hopelessly destroyed." The idea that the only things in the heavens are stars and planets would not fade quietly into the night.

However, Rosse did discover something about nebulae that would eventually play an important role in our understanding of them. With the Leviathan, Rosse clearly saw a spiral aspect in some nebulae. Of course, now we know that the spiral nebulae are galaxies like our own, but too far away to see individual stars in them with the Leviathan.

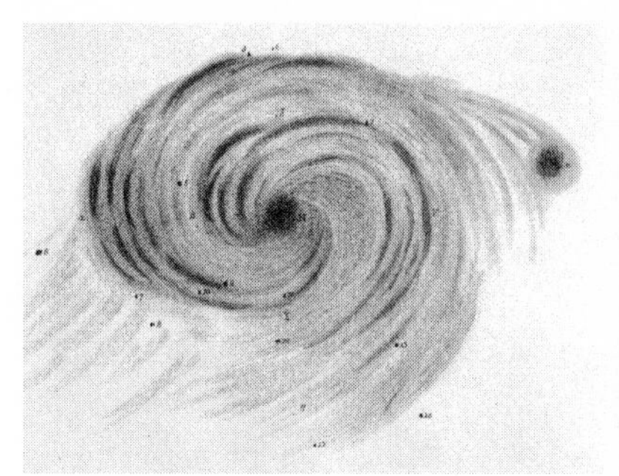

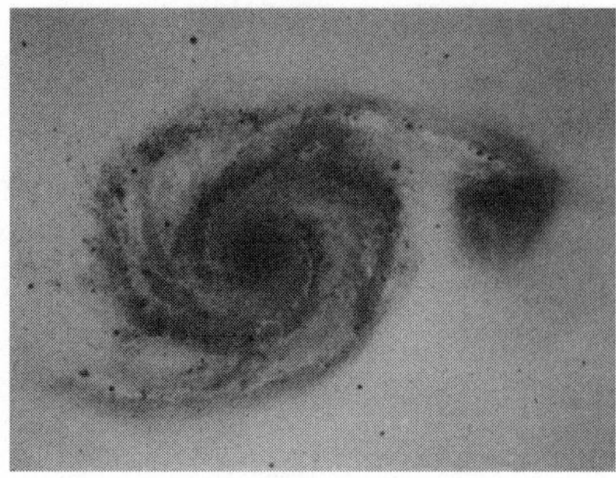

Using the Leviathan, in the spring of 1845 Lord Rosse discovered the spiral arrangement of the fifty-first entry of Messier's catalog. By 1850 he announced this discovery, along with fourteen other spiral nebulae (some of them later were shown to be spurious). On the left is Rosse's drawing of Messier 51, and on the right is a modern photograph.

Before following the thread of spiral nebulae, note how the issue of the ultimate resolvability of nebulae was settled. Although the use of large telescopes was (and still is) instrumental in the study of nebulae, the resolution came not from a monster telescope but from a rather small eight-inch telescope fitted with a new tool, the spectroscope.

In 1864 Sir William Huggins placed a crude spectrograph at the eyepiece of his telescope and discovered that while some nebulae have stellar spectra, a few have spectra that can be produced only by a luminous gas, heated by some external source. Thus, more than 250 years after Galileo first trained his optic tube on nebulae, we finally knew that while open clusters and globular clusters were made of stars, there were also other objects in the heavens, like planetary nebulae and the great nebula of Orion, which consisted of, in the words of Huggins, "a fiery mist...a shining fluid of a nature unknown to us."

But light from the great Andromeda nebula, and the other spiral nebulae as well, was too feeble for Huggins to tell if it was from starlight or a fiery mist. Again in Huggins's words, "its extreme feebleness made it uncertain whether the irregularities were due to certain parts enhanced by bright lines, or the other parts enfeebled by dark lines." Although one issue that had been debated for centuries was settled, as so often in the history of science, a larger issue loomed on the horizon: what are the spiral nebulae?[9]

THE GREAT ISSUE

Nebulae of a spiral nature were first discovered by Rosse with his giant telescope. But soon after Rosse finished the Leviathan, the character of astronomy was changed by the introduction of photography and spectroscopy, allowing much smaller telescopes entry into the nebula business. Photography opened another era in the nebula game. With the aid of photographic film, the light from dim nebulae could be concentrated during long exposures, allowing much smaller telescopes to freeze forever

[9] Not all galaxies are spirals; some are ellipticals, and some are neither and are called irregular. For instance, of the thirty-four galaxies in the Messier catalog, twenty-five are spirals, eight are elliptical, and one is irregular. Since the spiral galaxies were so strikingly different in appearance from other nebulae, the island universe debate centered on the nature of the spirals.

on film the images that before only the largest telescopes could fleetingly reveal to the eye. Early pictures confirmed in black and white the spiral structures Rosse saw with his eyes.

Although spectroscopy had settled the issue of the nebulous nature of planetary nebulae and diffuse nebulae, the spectra from spiral nebulae seemed to be unlike any stellar spectrum known at the time, but not quite like that from a diffuse nebula such as Orion.

The nature of spiral nebulae quickly settled to a choice between two hypotheses. Most astronomers thought that spiral nebulae were swirling gases in the process of forming stars and their concomitant planetary systems. Those who held that the spirals were protostars believed that they were relatively nearby.

A few astronomers believed that spiral nebulae were distant "island universes," as they were called by Immanuel Kant in his prescient 1755 book. Kant proposed that since nebulae were elliptical in

Different types of galaxies. The upper left is Messier 100, a nearby spiral galaxy. Upper right is a giant elliptical galaxy, Messier 87. Lower left is the one irregular galaxy in Messier's catalog, his eighty-second entry.

shape, like the Milky Way, they were external galaxies similar to ours.[10] Astronomers of the island-universe persuasion assumed that the spirals were very distant so that individual stars could not be resolved by even the most powerful telescopes of the time.

It would not be a fair representation to characterize the issue as a controversy, for almost all astronomers believed in the protostar theory. A sample of comments about the island-universe hypothesis conveys the general sentiment at the end of the nineteenth and beginning of the twentieth centuries:

- In 1888 the British astronomer Isaac Roberts produced the best photograph of the Andromeda nebula that had ever been taken. What every student of astronomy today immediately recognizes as a distant galaxy containing billions of stars, Roberts took to be an unmistakable picture of a nearby solitary star in the process of formation. He went on to associate two small nebulae near the main nebula (which are small, satellite galaxies of the Andromeda Galaxy) as planets in the process of formation around the central protostar.
- "No competent thinker, with the whole of the available evidence before him, can now, it is safe to say, maintain any single nebula to be a star system of co-ordinate rank with [that is, as large as] the Milky Way," wrote Agnes Clerke in the influential astronomy book *The System of the Stars*, 2d ed. (London: Adams and Charles Black, 1905).
- "Distant, isolated Milky Ways have never been sighted by man," declared astronomer Max Wolf in 1907, as quoted by S. L. Jaki in *The Milky Way* (New York: Science History, 1972).
- In 1907 the Swedish astronomer Karl Bohlin claimed that he could measure a parallax for the Andromeda nebula, and reported that it was at a distance of nineteen light-years (correct value: 2,120,000 light years).

Even in the twentieth century, as telescopes and photography improved, and the quality of the images of spirals approached what we

[10] While Kant was correct that our galaxy has a flattened, disk shape, his assertion that other nebulae are elliptical was based on the early catalog of Derham. But in his list of twenty-one nebulae, Derham characterized only five nebulae as elliptical, and we now know that only one of them is. A far-sighted theory was based on quite a shaky observational foundation.

have today, people continued to see what they expected to see. Just as Leonardo saw small holes in the septum when he held a heart in his hand, when astronomers held pictures of the Andromeda nebula to the light and pointed to the small points on the image we see as stars, they said that whatever they were, they certainly weren't stars. They claimed that the points of light on pictures of spirals were "too soft" to be stars. Today, when shown the same pictures, most twelve-year-old children will point out as stars the very objects professional astronomers judged were not. It would be easy to deride as blind watchers of the sky those who did not see, if not for the knowledge that we surely are also.

Rather than relate the many discoveries that either shed light or cast shadows on the island-universe debate, let us freeze a moment in time. An opportune occasion occurred on April 26, 1920, when two established astronomers, Harlow Shapley and Heber Curtis, appeared before the National Academy of Sciences in Washington to debate the possibility of island universes.

THE GREAT DEBATE

At the time of the debate, Harlow Shapley was an astronomer at the Mount Wilson Observatory in California, and later became director of the Harvard Observatory. He was best known for the proof that our solar system is not located near the center of the Milky Way. His position in the debate was that our galaxy is much larger than was commonly believed at the time, and encompassed the spiral nebulae.

Heber Curtis was director of the Allegheny Observatory near Pittsburgh, a distinguished astronomer who devoted his life to photographic and spectroscopic study of nebulae. He later served as director of the observatory at the University of Michigan. In the debate he defended the position that the extent of the Milky Way was much less than Shapley proposed, and that the spiral nebulae were distant external galaxies, equal in rank to the Milky Way.

The part of the debate related to the island-universe issue hinged on the interpretation of six agreed-on pieces of observational evidence:

1. Ardiaan van Maanen, Shapley's colleague at Mount Wilson, claimed that examination of pictures of the spiral nebula Messier 101 showed evidence of rotation. In pictures taken years apart he found a displacement of the stars in the spiral. This observation was confirmed by Dutch and Russian astronomers. If the nebula

Participants in the great debate on the nature of spiral nebulae and the size of the Milky Way were Harlow Shapley (left) and Heber Curtis (right).

was a separate star system at a distance required by the island-universe hypothesis, the stars of the galaxy would have to be moving faster than the velocity of light to have a detectable displacement. (This observation, as well as the confirming observations, was later shown to be spurious, the result of astronomers performing a difficult observation at the limit of the resolution of their instruments, and discovering what was placed on the photographic plates by their minds, not by nature.)

2. The sudden increase in the brightness of objects in spiral nebulae had been noted and studied for decades. If these were the common (if forty times per year in our galaxy can be considered common) abrupt rise in brightness of stars called nova, then the nebulae would have to be outside our galaxy for the nova to appear so faint. However, in 1885 an object, which appeared to be in Andromeda, had dramatically increased in brightness, and for a brief period became nearly as bright as the entire rest of the nebula combined. If this was indeed the brightening of a single star, and the nebula was as distant as the island-universe hypothesis required, then the star must have been as bright as a billion suns. (It was indeed! The spectacular event of 1885 was a supernova in the Andromeda galaxy, the same type of phenomenon that occurs once every century or so in our galaxy. The "new star" of Tycho Brahe was a supernova in our galaxy. The last visible galactic supernova was witnessed by Johannes Kepler.)

3. Although it could not be stated that the spectra of spiral nebulae were as simple as the spectra from diffuse nebulae like Orion, neither did the spectra of spiral nebulae resemble the spectra of any of the hundreds of well-studied stars in our solar neighborhood. The opponents of island universes seized on this and said that if spiral nebulae consisted of stars, they should have stellar spectra. (Now we know the spectra of spiral nebulae are dominated by the light from giant stars, which are very rare in the neighborhood of the Sun.)

4. At the time of the great debate, the locations in the sky of about fifteen thousand nebulae were known. They were not uniformly distributed in all regions of the celestial sphere, but seemed to avoid the region of the plane of the Milky Way. (Now we know that we can't see through the Milky Way because of the large amount of dust in the plane of our galaxy. Therefore, we don't see the extragalactic nebulae in the direction of our galactic plane.)

5. We can count the number of stars we see around us and imagine how bright the light from these stars would appear if viewed from outside our galaxy. The spiral nebulae appear brighter than our galaxy would if viewed from the distances assumed by the island-universe proponents. (Now we know that the density of stars is smaller in our neighborhood than in the center of our galaxy, but the light from spiral nebulae is dominated by the light from the relatively dense inner core.)

6. Several of the spiral nebulae seemed to be traveling away from the center of our galaxy at high speed, between five hundred and two thousand miles per second. (No one at that time knew about the expansion of the universe.)

In the course of the debate, Curtis and Shapley tried with varying degrees of success to interpret the observations to fit their hypotheses. Both were experienced enough to realize that not all observations at the leading edge of science turn out to be correct. The art of science is knowing which observations to ignore and which are the key to the puzzle.

We can summarize the great debate by quoting from the concluding remarks of the participants, first from those of Shapley:

> Another consequence of the conclusion that the galactic system is of the order of 300,000 light-years in greatest diameter, is the previously mentioned difficulty it gives to the "comparable galaxy" [island universe] theory of spiral nebulae. I shall not undertake a description and discussion of this debatable problem. Since the theory probably stands or falls with the hypothesis of a small galactic system, there is little point in discussing other material on the subject, especially in view of the recently measured rotations of spiral nebulae [by van Maanen] which appear fatal to such an interpretation.
>
> It seems to me that the evidence, other than the admittedly critical tests depending on the size of the galaxy, is opposed to the view that the spirals are galaxies of stars comparable with our own. In fact, there appears as yet no reason for modifying the tentative hypothesis that the spirals are not composed of typical stars at all, but are truly nebulous objects. Three very recent results are, I believe, distinctly serious for the theory that spiral nebulae are comparable galaxies—(1) Seares' deduction that none of the known spiral nebulae has a surface brightness as small as that of our galaxy [point 5 above]; (2) Reynold's study of the distribution of light and color in typical spirals, from which he concludes they cannot be stellar systems [point 3 above]; and (3) van Maanen's recent measures of rotation in the spiral M[essier] 33, corroborating his earlier work on Messier 101 and 81, and indicating that these bright spirals cannot reasonably be the excessively distant objects required by the theory [point 1 above].

Curtis closes his remarks with these words:

> There is a unity and an internal agreement in the features of the island universe theory which appeals very strongly to me. The evidence with regard to the dimensions of the galaxy, on both sides, is too uncertain as yet to permit of any dogmatic pronouncements. There are many points of diffi-

> culty in either theory of galactic dimensions, and it is doubtless true that many will prefer to suspend judgment until much additional evidence is forthcoming. Until more definite evidence to the contrary is available, however, I feel that the evidence for the smaller and commonly accepted galactic dimensions is still the stronger; and that the postulated diameter of 300,000 light-years [by Shapley] must quite certainly be divided by five, and perhaps by ten.
>
> I hold, therefore, to the belief that the galaxy is probably not more than 30,000 light-years in diameter; that the spirals are not intra-galactic objects but island universes, like our own galaxy, and that the spirals, as external galaxies, indicate to us a greater universe into which we may penetrate to distances of ten-million to a hundred-million light-years.

Although such scientific debates are often fun to witness, they never settle an issue. It is even more difficult to judge who "won" the debate, because in science, nature—not a panel awarding debating points—is the ultimate arbitrator.

Shapley was closer in his estimate for size of the Milky Way, the major point he was trying to convey, and Curtis was right about the island universes, his main subject.

Based on the scientific evidence known at the time of the debate, an unbiased observer probably would have judged Shapley the winner on the island-universe part of the debate.[11] Let's briefly recap the main points in the debate as a modern-day television commentator might with the help of a panel of experts.

1. Motion of Messier 101

COMMENTATOR: It seems to me that the motion detected by van Maanen (and confirmed by two others) was striking evidence in support of Shapley's position. Curtis could only say that the observations must be wrong.

PANEL: It was not very scientific for Curtis to dismiss the observations; point 1 to Shapley.[12]

[11] This illustrates why unbiased scientists rarely make significant advances.

[12] Curtis turned out to be right.

2. Variable Stars

COMMENTATOR: I think that here the evidence is not as clear-cut. Although some novae in Andromeda seemed to be consistent with Curtis's estimate for the distance, the unprecedented brightness of the 1885 event strongly suggested that Andromeda must be closer than Curtis requires. Curtis could only answer that perhaps the event of 1885 was a wholly new class of phenomenon.

PANEL: Curtis was not very convincing in inventing a new phenomenon; point 2 to Shapley.[13]

3. Spectra of Spirals

COMMENTATOR: Shapley simply said that the fact that the spectra and color of the spirals do not resemble well-studied stellar spectra proved the point that they were not made of stars. Curtis had to postulate that there were two types of stellar populations, one type in spiral arms like those near us and another near the center of galaxies, which had a different type of spectrum and dominated the light from the spirals.

PANEL: Sounds like too much postulating by Curtis; point 3 to Shapley.[14]

4. Spatial Distribution

COMMENTATOR: Shapley pointed out that spirals are not seen toward the plane of the Milky Way, which must be proof that they somehow know about the orientation of our galaxy and must therefore be part of it. Curtis's reply was that some spirals seem to be filled with some material that obscures starlight. *If* our galaxy has such a band of obscuring material in its center, and *if* the Sun is very near the center of the plane of the Milky Way, then it would obscure our vision out of our galaxy toward the plane of our galaxy.

PANEL: It is not good form for Curtis to make two assumptions to explain one fact; point 4 to Shapley.[15]

[13] Curtis turned out to be right.

[14] Curtis turned out to be right.

[15] Curtis turned out to be right.

5. Brightness of the Nebulae

COMMENTATOR: Shapley simply reiterated that once again, spiral nebulae do not resemble our galaxy. Curtis had no ready reply.

PANEL: Curtis is on the ropes; point 5 to Shapley.[16]

6. Velocities

COMMENTATOR: Neither Shapley nor Curtis had a clue about the reason for the large recessional velocities of the spirals.[17]

PANEL: Both sides are exhausted at the end of the debate; point 6 is a tie.

7. Concluding Remarks

COMMENTATOR: Shapley's concluding remarks were full of facts, whereas Curtis seemed soft on the issues.

PANEL: Scientists should stick to the facts; final point to Shapley.

Based on a fair judgment of the facts as presented in the debate, Shapley was the winner in the debate with Curtis on the cards of all members of the panel by a score of six victories, no defeats, and one tie.

Less than five years after the great debate, Edwin Hubble proved beyond a doubt that Curtis was right about the nature of the spirals.

Just because Curtis was right doesn't mean that he was a better scientist. Curtis and Shapley both viewed the same body of evidence and came to different conclusions. Although Shapley's marshaling of facts at the time *seemed* to result in a stronger case than that presented by Curtis, some of the facts were wrong, and some of them were incorrectly interpreted.

The Shapley-Curtis debate illustrates the difference between the scientific method and the process of science. Ultimately experiment and observation is the arbitrator in the confrontation of theory and nature, but the process of science does not always involve carefully and impartially weighing the evidence. Science does not proceed like a cookbook recipe in the making of a hypothesis, comparing its prediction with observations and either accepting or rejecting the hypothesis. There is always confusion at the leading edge of research, and there are

[16] Curtis turned out to be right.

[17] They were unaware of the new cosmological developments making use of Einstein's new theory of gravity.

always a few discrepant and contradictory pieces of information that can't be explained.

Rereading Curtis's concluding statement, I am struck by how soft and squishy his arguments seem compared with Shapley's. Yet Curtis was right. It is not enough to be convincing; you have to be right.

THUNDER FROM THE SKY

On October 6, 1923, Edwin Hubble opened the dome of the giant one-hundred-inch telescope on Mount Wilson in the foothills of the San Gabriel mountains outside of Pasadena, California, and, in the clear pre-smog, pre-urban-sprawl southern California night, turned the telescope in the direction of an outer spiral arm of the Andromeda nebula. He exposed a sensitive photographic plate for about a half hour to the light from Andromeda, and after developing the plate, detected among the dense swarm of images the unmistakable image of a star.

It was not an ordinary star like the Sun, but a type known as a Cepheid variable star. The apparent brightness of Cepheid variable stars is not constant; as advertised, they vary, becoming brighter and dimmer with time in a predictable fashion. In 1912 the astronomer Henrietta Leavitt had made an exhaustive study of Cepheid variable stars in our galaxy and in the Magellanic Clouds and discovered that while the interval of time for the periodic brightening and dimming ranges from a couple of days to a couple of months for different stars, there is a definite relationship between the intrinsic luminosity of the star and how long it takes the star to brighten and then fade.

Hubble knew that if he could measure the time it takes for the star in the Andromeda nebula to brighten and fade, he could learn the intrinsic luminosity of the star. Then by using the **inverse-square** law for the decrease of brightness with distance, he would know how far away the star was, as well as the nebula that contained it.

Of course, Leavitt could determine the intrinsic luminosity of Cepheid variable stars because she knew the distance to them. To know that, she had to understand the distance scale in our galaxy as well as the distance to the Magellanic Clouds. This required the previous steps on the distance ladder, such as the Hertzsprung—Russell diagram, which was established because we knew the distance to nearby stars by their parallax shift, which was known in terms of the Earth–Sun distance, which was first determined in terms of the Earth–Moon distance, which was first measured in terms of the radius of Earth, which Eratosthenes

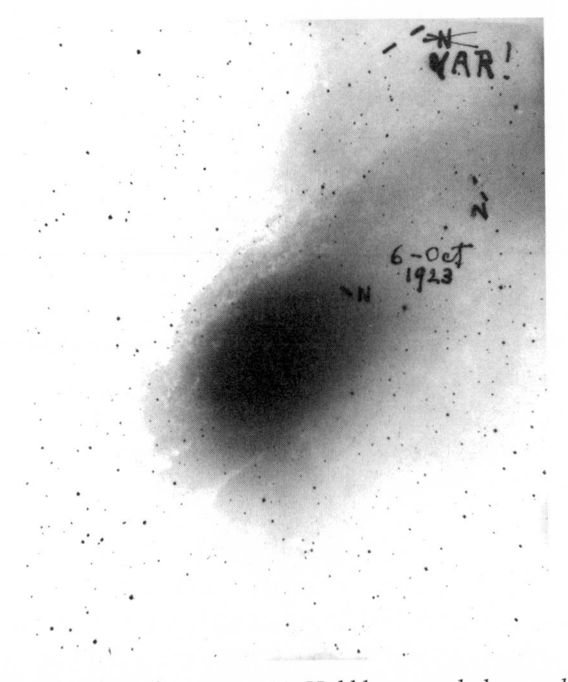

Over the night of October 5–6, 1923, Hubble exposed photgraphic plate number H325H (the 325th photograph taken by Hubble with the 100-inch Hooker telescope) to the light from the Andromeda nebula. After developing the plate he noticed three unfamiliar bright dots. He marked the three with an "N" thinking were probably ordinary novae, stars that brighten once then remain dim. But after checking his previous photographs of Andromeda he realized that one of the stars had brightened and dimmed before, and must be a variable star rather than a nova star. Then he crossed out the N and wrote "VAR!" to indicate the position of a variable. Hubble then knew he had the ruler to measure the distance to the nebula.

measured using a hole in the ground in Aswan. Never has a simple hole in the ground led to such a journey![18]

Over the next year Hubble found additional Cepheid variable stars in Andromeda, as well as a number of them in a second nearby spiral nebula. After exposing 130 photographic plates to the light from Andromeda and another 65 plates to the light from the other spiral nebula, Hubble had collected enough information about a sufficient

[18] Whenever I hear the expression that someone doesn't know something from a hole in the ground, I think about this.

number of Cepheid variable stars in the two nebulae to measure their periods and determine their intrinsic luminosity.

By the end of the year 1924, there was one person in the world who knew the distance to Andromeda, one person who knew that it was indeed an island universe, well outside our own galaxy. Among the billions of inhabitants on a small planet orbiting about a typical star nestled in an outer spiral arm of an average galaxy, there was one individual who knew for sure of the immense vastness of the universe, and glimpsed the enormity of the multitude of galaxies in the universe.

On the last day of the year 1924, Hubble cabled a copy of a paper containing his results to fellow astronomer Henry Norris Russell to be read the next day during a New Year's Day session of a joint meeting of the American Astronomical Society and the American Association for the Advancement of Science in Washington, D.C. And then the entire world knew what only Hubble had known.

On New Year's Day 1925, while Hubble sat at home in Pasadena puffing on his ever-present pipe, he heard the unmistakable rumble of distant cheering in appreciation for a mighty triumph. The cheers came not from Washington, where Russell was reporting Hubble's results for twelve Cepheid variable stars in Andromeda and twenty-two Cepheid variables in the other spiral nebula to a quiet, attentive audience of less than a hundred astronomers, but from the nearby Rose Bowl in Pasadena, where a football team from Notre Dame led by "four horsemen" was playing a Stanford team coached by Glenn "Pop" Warner to the delight of a sellout crowd of fifty-three thousand screaming fans.

Although the cheering Hubble heard on that momentous day when the universe changed were "volley cheers on high, shaking down thunder from the sky" as fans sang the Notre Dame fight song while the Fighting Irish won 27–10, he knew that a hundred astronomers on the other side of the country truly appreciated that from that day onward, we would view the universe in a different way.

While the coach who guided Notre Dame to victory was carried off the field at the end of that day, adding to the growing legend of someone by the name of Knute Rockne, Harlow Shapley and Heber Curtis had a drink together after the session in Washington and discussed how the three-hundred-year-old debate had been settled by another growing legend—Edwin Hubble. They knew a new era in astronomy had begun. Shapley and Curtis, more than others, appreciated that in centuries and millennia to come, it would be the triumph of the quiet man in Pasadena that would be celebrated as long as our species has the curiosity to watch the sky.

Unlike most advances in science, this one was immediately appreciated as one of the greatest discoveries of the century. As appreciation for opening a new chapter in astronomy and extending the horizons of human investigation beyond the confines of our galaxy into the distant reaches of the universe, Hubble received one of the $500 awards given by the American Association for the Advancement of Science to the top two papers presented at the meeting. Hubble shared the limelight and the prize money with the other winner, the author of a paper on the digestive tracts of termites.

PART III

The Universe

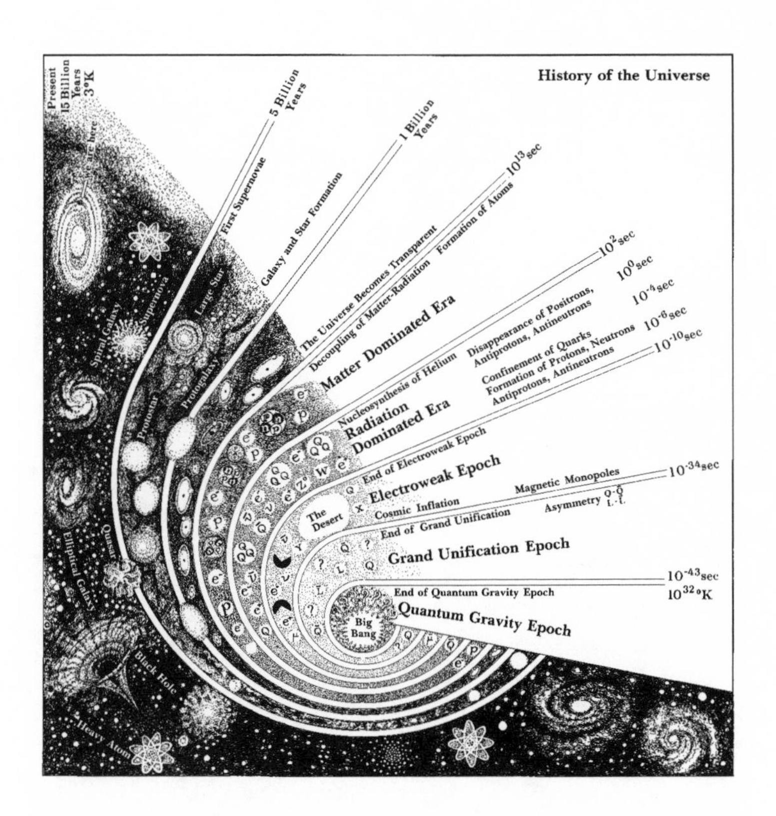

EIGHT

The Expanding Fog

When teaching modern cosmology to graduate students at the University of Chicago, I always stroll into the classroom on the first day of the term (fashionably late) and, without saying a word, immediately proceed to write an equation on the blackboard. No use fooling around, it's best to get right down to business, because the universe is a big place to cover in only one semester of lectures. In fact, sometimes I can cover only a small part of it before the end of the term and have to leave large regions of it unexplained. I get right down to business because students at Chicago are a serious lot, ready to work, and by the time they are advanced graduate students they have already heard most of my jokes anyway. After writing this particular equation in large, bold symbols and drawing a box around it, I just have to stand back for a minute and admire it in silence before proceeding.

As a theoretical physicist I am one of those odd people who can see beauty in a mathematical equation. I am not odd enough to think that just any old mathematical equation is beautiful; those *really* weird people who think that all equations are beautiful are known as mathematicians. No, I don't like all equations, for some leave me with a numb, empty feeling inside, like one has after listening to a Barry Manilow song. But other equations give me great pleasure when I simply write them on a blackboard. The ones I enjoy are not just mathematical equations, but mathematical equations that say something about nature.[1]

One can represent nature in many ways. A Joseph M. W. Turner painting of a storm-tossed sea is a depiction of the beauty and power of the wind and waves, but so are the equations of hydrodynamics. The painting can be appreciated by a wider audience, but the equations can express something that can't be put on canvas. Mastery, or at least an understanding, of the equations gives us poor, pitiful humans some degree of comprehension of the forces of nature. What a wonderful world, where we can enjoy both an artistic interpretation and a mathematical description of the same scene!

After admiring the equation just long enough for the students to start to wonder if I have been inhaling too much chalk dust in my office, I turn to them and say, "I assume you are familiar with **Einstein's equations** from his theory of gravity." Naturally they all nod yes (even if they have never seen the equations before). I then proceed in two quick minutes to demonstrate that Einstein's theory of gravity does not describe a universe that is unchanging in time but one that is either expanding or contracting. The students nod again. Of course, then I remind them that we will be particularly interested in the expanding solution, because that's the one that seems to describe the universe in which we live.

Even skeptical students readily accept the argument, because, damn it, an expanding (or contracting) universe *is* an implication of Einstein's theory. Today when we look at the equations, expansion sort of screams out at us. But for nearly fifteen years after Einstein developed his theory, it was not so easy to hear what the equations were saying.

In *On War*, Carl Philipp Gottlieb von Clausewitz talks of the "fog of war," where the unfolding of events on the field in the heat of battle are not as clear as they are later to armchair generals, and contributing to the outcome of a battle are psychological and accidental factors that elude exact calculation. Perhaps when describing to the class how naturally the expanding universe follows from Einstein's theory I am playing the role of armchair cosmologist from which I can see all the pieces of

[1] Mathematicians say that all equations describe nature and are beautiful, but like one of those three-dimensional posters where you just can't find the image no matter how hard you squint, I can't see the beauty. I am envious that mathematicians can appreciate things I cannot.

the problem neatly laid out on the board. But the idea of an expanding universe didn't follow immediately as a consequence of Einstein's equations when people first looked at them. At the cutting edge of cosmology, in fact of any branch of physics, things never appear quite so clear as they do later. Most scientists spend their lives groping around trying to find their way, as if lost in a fog.

THE FOG OF PHYSICS

At the end of World War I there were several people in different countries lost in the fog of physics trying to apply Einstein's new theory of gravity to cosmological problems. Although many realized that Einstein's theory was a powerful conceptual tool, cosmology was a new arena in which to apply the ideas, and it was not at all clear how to fashion a new view of the universe from the theory. Rather than use Einstein's theory of gravity (known as the general theory of relativity, or **general relativity** for short) as a precision instrument to fashion a new cosmology, they first employed it as a crude hammer to pound down some rough edges in Newtonian cosmology. It was as if the physicists lost in the fog were mechanically inclined people of the Middle Ages given a flashlight and told to use it to produce illumination, so they went ahead and set fire to it.

There was a surprising lack of communication between cosmologists. They passed each other lost in the fog, largely unaware that others were wandering in the thick mist of confusion. In the battle to expose nature's secrets, they didn't know the location of the enemy; they didn't know the identity of the foe, or even if there was one to be found. It seems that the only thing they all felt were the tremors from Einstein's new theory of gravity.

Just as Newton's theory burst on the world in near final form in the presentation of the *Principia* to the Royal Society of London, Einstein's theory of gravity was unveiled to the Prussian Academy of Science on November 25, 1915. Of course, the theory did not come to Einstein while dreaming under an apple tree, but as the result of eight years of long, hard struggle after discovering the principle on which the theory was based in 1907.

Even before the theory received widespread acceptance, Albert Einstein turned his attention to its cosmological implications in a 1917 paper, "Cosmological Considerations on the General Theory of

Relativity." In the fog of physics, Einstein was searching for two things from the cosmological applications of his theory:

1. a universe that is unchanging in time, or *static;* and
2. a universe in which space doesn't exist without the presence of matter.

Newton's theory of gravity did not satisfy either of these conditions, and Einstein thought that his grander edifice might.

The first condition is simply a prejudice that reflected the worldview of Einstein, indeed of many who wondered about the universe. Whether they believed that the universe had existed forever or was created in a special event (say, requiring six days of effort followed by a day of rest), they typically assumed that as long as it had existed, it had more or less the appearance it does now—it was static. Sure, things move around a bit, the planets orbit the Sun and stars seem to have small motions through our galaxy, but it was thought that the motions were insignificant, and the universe today looks pretty much like it always did. But this was seven years before Hubble demonstrated that galaxies distinct from our own exist, and twelve years before he discovered that distant galaxies recede from us. In some sense Einstein worked within the confines of a smaller universe than we do now.

Newton's universe lacked the stability sought by Einstein. In Newton's theory every mass attracts every other mass with a force decreasing according to the **inverse-square** law. Therefore, if you imagine an infinite space filled with stars, every star would feel the tug of every other star. If the stars are placed on a perfectly ordered latticework, with every star in its particular assigned place, then it is possible to have a perfectly symmetrical arrangement where every star will be pulled equally in all directions and will feel no net force in any one direction.

However, if the stars are not tied down to the latticework and are allowed freedom of movement, then any small departure from the initial symmetrical arrangement will destroy the balance and lead to disaster. The stars will no longer be pulled equally in all directions and will eventually start to collapse into tight knots. In turn, the tight, dense knots will suck in even more matter, leading to a catastrophe. Newton realized that his universe could not be everlasting, and imagined it was God's duty to roll up her sleeves after the inevitable collapse and wade in to separate the stars and put them back in their proper places, much as parents periodically have to step in and separate squabbling siblings.

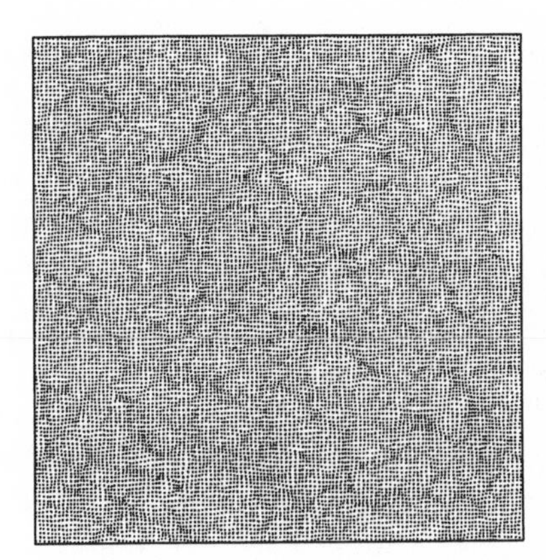

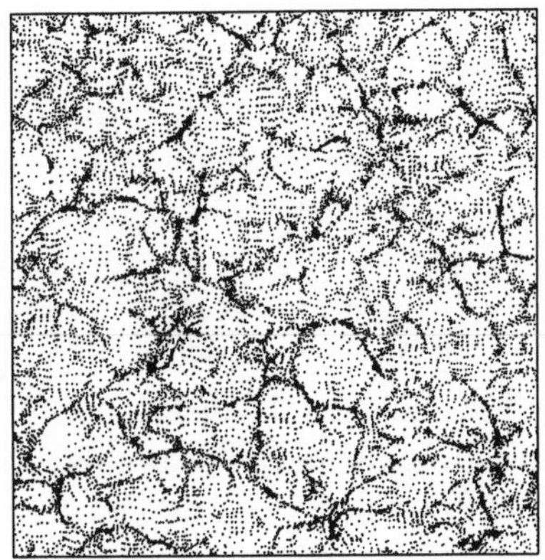

These two panels illustrate how in Newton's theory of gravity, small departures from absolute uniformity will lead to the formation of tight, dense knots. Imagine that stars are distributed like the black dots in the left panel—spread out roughly, but not exactly, *evenly in space. After some time has passed, regions that started with a slightly higher density of mass attract surrounding masses, and the nonuniformity grows in time as shown in the panel on the right. Eventually all the masses will collapse into a single, large conglomeration.*

The second condition Einstein imposed on his cosmological solutions was a reaction against the concepts of "absolute space" and "absolute time" that lie at the foundations of Newton's dynamics. As Newton said in the *Principia*,

> Absolute space, in its own nature, without relation to anything external, remains always similar and immovable.
>
> Absolute, true, and mathematical time, of itself, and from its own nature, flows equably without relation to anything external.

In Newtonian dynamics, matter is just a player on the fixed and immutable field of space and time, playing the game according to the rules of classical dynamics. But in Einstein's picture space and time are allowed to participate in the game as well, and in the interplay among space, time, and matter, space and time are not absolute, but, well, relative.

Einstein was not the first to question the fixity of space and time. Ernst Mach, a German mathematician, physicist, and philosopher, had suggested that rather than describing the motion of a body with respect to absolute space, motion makes sense only relative to an idealized system of distant, fixed stars. Thousands of papers have explained, refuted, or expounded on Mach's principle, written by physicists, philosophers, and other assorted characters (including a cosmologist wannabe named Vladimir Ilych Ulyanov Lenin). Machian ideas guided Einstein in the development of his theory of gravity, but he later refuted the principles behind them. In 1917, however, Einstein was a Machian, and he believed that space itself could not exist without the presence of matter.

So with the twin goals of finding a static universe and restricting space to exist only in the presence of matter, Einstein turned to cosmology armed with his new theory of gravity. Although he had the right weapon in his hands, in the fog of physics he was firing at the wrong targets.

Einstein first tried to find solutions to his equations that would describe a universe infinite in space that satisfied his two desiderata, but he quickly realized that such a solution could not exist. Then he tried to find the desired solution in a universe of finite spatial extent, again failing. Einstein found that a universe containing matter and radiation just didn't want to stay put, but naturally seemed to want to expand or contract. (You just can't keep a good universe down.) Rather than question the original goals he set for his cosmological model in light of what he found, Einstein chose to butcher the theory he had spent the previous seven years perfecting. As he wrote in his first paper on cosmology:

> I shall conduct the reader over the road that I have myself travelled, rather a rough and winding road, because otherwise I cannot hope that he will take much interest in the result at the end of the journey. The conclusion I shall arrive at is that the field equations of gravitation which I have championed hitherto still need a slight modification.

The "slight modification" consisted of adding to his beautiful equations an ugly new term, whose sole purpose was to keep the universe static. If you are not one of those odd characters who can see sublime beauty in a mathematical equation, try to imagine how Einstein's slight modification destroyed the coherence of his equations by picturing Leonardo returning to the Mona Lisa two years after painting it and deciding to modify it slightly by adding a mustache. The ugly term Einstein added is known as the cosmological term, and it can be adjusted in such a way as to balance the effect of matter and radiation that tends to drive expansion. Fifteen years later, after the discovery of the expansion of the universe, Einstein referred to this modification as "the biggest blunder of my life."

Einstein's paper has caused more than its share of consternation among historians and physicists. Until this time Einstein's instincts as a physicist rarely led him astray. His intuition and insight had been uncanny. But here he is truly lost in the fog like a normal human physicist. It is sometimes said that Einstein missed the opportunity to predict the expansion of the universe twelve years before Hubble's discovery. Of course, in 1917 the extent of our own galaxy was not even known, and the existence of other galaxies was not given much credence. Perhaps it is too much to expect *anyone* to have predicted the expansion of the universe. Einstein was not just *anyone*, however, and although we do not *sense* the expansion of the universe, I believe it was possible for Einstein to *reason* it.

In 1905 Einstein looked at the equations written on the papers spread across his desk in the Bern patent office and realized that years of toil led him to the unanticipated conclusion that space and time were not independent, but were interrelated in a deep way. Somehow, through the sheer effort of *reason conquering sense*, he was able to reject the assumed independence of space and time that had been a part of our worldview for longer than history can chronicle, and imagine a unified space-time.

Ten years after this remarkable leap of imagination, after many additional years of struggle, he realized that his equations implied that space was not flat, but curved in the presence of matter. Again, through the remarkable process of *reason conquering sense*, he was able to dismiss the flat-space geometry of Euclid that had ruled physics and cosmology for two millennia and embrace curvature of space.

Then, two years later, he discovered that the theory he had spent years of his life and almost unimaginable intellectual effort developing did not admit solutions describing the universe as he sensed it must be. And so in 1917, the reason of the man who could change his perception of space and time, and who could modify his picture of the geometric construction of space, finally succumbed to sense, for when he looked at his equations in light of a preconceived sense of the universe, he said they "still need a slight modification," because we do not sense the expansion of the universe.

Whether or not we realize it, perhaps our ingrained sense of the universe is more fundamental than our sense of space, time, or geometry, and drive even the greatest among us to be, on occasion, a blind watcher of the sky.

Einstein was not alone in the fog, for shortly after he published his first paper on the cosmological implications of his new theory of gravity, a Dutch astronomer, Wilhelm de Sitter, also wandered into the fog. Like Einstein, de Sitter assumed a universe that was unchanging in time, but he took a different view of the meaning of "unchanging" from that of Einstein. Whereas Einstein assumed a static universe, where nothing moved except for the inconsequential motion of the stars, de Sitter allowed the universe to expand, but with the provision that the rate of expansion must be the same for all time.[2] Although in the de Sitter model the universe expands, it always expands with the same rate, and so in some sense it looks the same for all time. In the big-bang model, which was developed *after* de Sitter's model, the expansion rate changes as the universe expands, and the universe began a finite time in the past. So de Sitter's cosmology is not truly a big-bang model because it has no beginning, and the universe always has essentially the same properties, independent of time.

De Sitter discovered this curious solution by including Einstein's cosmological term, *but throwing away matter and radiation*. This seems rather an ironic twist, because Einstein added the cosmological term to prevent the expansion (or collapse) driven by matter in the universe. Now de Sitter removed matter and kept the cosmological term and found expansion caused by the very term introduced to prevent it. Ah, the fog was thick.

All the while the equations were screaming out for an expanding solution that changes with time, and Einstein and de Sitter were going through incredible contortions not to discover it. However, de Sitter's model shared one property with what would eventually become the big-bang model: the red shift of the spectral lines from distant sources. As de Sitter noted in his paper,

> the lines in the spectra of very distant stars or nebula must therefore be systematically displaced towards the red, giving rise to a *spurious* [my emphasis] positive radial velocity.

This is exactly what Hubble would discover twelve years later. But the fog had not yet cleared. De Sitter's cosmological model could not be a realistic model for our present universe because it did not actually con-

[2] Technically, such a universe is called *stationary* rather than static.

tain matter or radiation. It was an empty universe with the only energy or mass in the cosmological term, so most people did not take it seriously as a description of the physical universe. He also used the unfortunate adjective *spurious* in describing the displacement of the spectral lines. However spurious he thought the displacement to be, he did discuss the very weak observational evidence about the spectra of nebulae that existed in 1917. Sometime later, the German physicist Hermann Weyl made it clear that in the de Sitter model the displacement of the spectral lines would increase with the distance to the source, again in anticipation of Hubble's discovery.

Slowly the fog was lifting, but the person who would first penetrate the mist and see the naturalness of the expanding universe had not yet even heard of Einstein's theory of gravity at the time Einstein and de Sitter were groping about. In fact, Aleksander Aleksandrovich Friedmann was not even a professional physicist at that time. In 1917, the twenty-two-year-old Friedmann, who was by training an atmospheric physicist with a degree in meteorology from the University of Saint Petersburg, was chief of a Russian factory manufacturing aviation instruments.

A METEOROLOGIST LIFTS THE FOG

Many people believe that all great ideas in physics are sparked by a crisis. It is usually said that Copernicus developed the heliocentric model simply because the old geocentric model was no longer able to predict correctly astronomical phenomena such as planetary conjunctions or eclipses. But in fact, Copernicus was motivated more by aesthetics than by the desire to produce a model for more accurate predictions. Aesthetics has guided many geniuses on the path to discovery. They seem to use it as their own special lamp to see things others do not.

Einstein was motivated by aesthetics in his development of relativity. Although there was a crisis in physics at the beginning of the twentieth century because of the Michelson-Morley experiment that proved the aether did not exist, it does not seem to have been the force that pushed Einstein toward the **special theory of relativity**. Surely after Einstein relegated the aether to the dustbin in 1905, there was no crisis in physics that demanded Newton's theory of gravity be modified. Rather, Einstein was led to the **general theory of relativity** only by the aesthetic requirement that gravity also conform to the ideas of relativity.

By 1907 Einstein started to apply his special theory of relativity to gravitational problems, and realized that the apparent position of a star would be shifted if the Sun is very close to our line of sight to the star. In 1911 he again returned to this phenomenon, usually referred to as bending of light, and calculated that a light ray from a distant star grazing the Sun would seem to be deflected by 0.83 seconds of arc. He immediately seized on the importance of the deflection of light by the Sun, and in his paper he urged astronomers to look for the effect. Although his reasoning was qualitatively correct, the full general theory of relativity was still in its embryonic stage, and the preliminary value of 0.83 seconds of arc in his paper was wrong.

Astronomers did indeed look for the effect. Because 0.83 seconds of arc is a small deflection, the image of the star would still appear to be very close to the Sun. You don't have to be an astronomer to know that stars are not easily seen in daylight, particularly in the direction of the Sun, so astronomers would have to look near the Sun when the sky was dark. This is possible only at the time of a total solar eclipse.

The first attempt to look for the effect was an Argentinean expedition in 1912 that attempted to observe a solar eclipse in Brazil. The sky was dark enough, but unfortunately it was also raining, and the Sun could not be seen through the clouds. Two years later the German industrialist (and weapons magnate) Gustav Krupp financed a German expedition into Crimea to look for the effect. But while the expedition was preparing for the eclipse, World War I broke out, and the German astronomers were arrested and detained by the Russian authorities. Eventually they were allowed to return home, but the clouds of war prevented any observation during the eclipse.

In November 1915, Einstein completed the general theory and realized that because of the special effect of the curvature of space, the bending of light by the Sun would be 1.7 minutes of arc, twice as large as his first estimate. Thus the bending of light was a way to detect whether space is curved, and a definitive discriminant between Einstein's theory of gravity and Newton's.

After the war there was a total solar eclipse on May 29, 1919. Two British expeditions were organized to search for the bending of light as it passed near the Sun—one to Sobral, Brazil, led by Andrew Crommelin, and one led by Arthur Stanley Eddington to the Portuguese Island of Principe near Spanish Guinea off the western coast of Africa.

On November 6, 1919, before a joint meeting of the Royal Society of London and the Royal Astronomical Society, Crommelin and Eddington reported results that confirmed Einstein's prediction for the bending of light, and sounded the death knell for Newton's theory as the final word on gravity.[3]

It was immediately realized that the world after that date was a different place. On November 7, on page 12 of the London *Times*, an article appeared under headlines in bold print proclaiming **REVOLUTION IN SCIENCE—NEW THEORY OF THE UNIVERSE—NEWTONIAN IDEAS OVERTHROWN—MOMENTOUS PRONOUNCEMENT—SPACE "WARPED."** News of the revolution in science soon spread across the world.

Eventually the news reached Saint Petersburg, where a young professor at the university there read of the hysteria. Caught up in the excitement, in a short time Aleksander Friedmann taught himself general relativity and turned his attention to cosmological problems.

While previous applications of Einstein's theory to cosmology seemed tentative and conservative, Friedmann saw that the most natural solutions describe an expanding universe (unlike Einstein's model) where the rate of expansion changed in time (unlike de Sitter's model). In Friedmann's model the universe was not infinitely old like in the models of Einstein and de Sitter, but had emerged at a finite time in the past.

Looking back as an armchair cosmologist it is easy to see that there were observational hints of a universal expansion that people might have picked on and made the connection between expansion and relativity. But Friedmann's development was not based on any experimental evidence or crisis. Instead, it came from the simplicity and aesthetics of the solution. Whereas Einstein, de Sitter, and others tried to distort Einstein's theory and make it conform to the preconceived notion of an infinitely old universe, Friedmann just did what the theory seemed to demand. Perhaps the fact that Friedmann was largely self-taught and worked in relative isolation allowed him to view the problem from a different perspective and helped him peer through the fog.

In July 1922 Friedmann published his theory of an expanding big-bang universe in a prestigious German physics journal. Einstein was quick to respond. He did not immediately embrace the expanding-universe cosmology, but in September 1922 he published a one-paragraph response to Friedmann's paper in which he claimed that Friedmann had

[3] One of the wonderful things about science is its international aspect. Less than one year after a bloody and bitter conflict, the theory of the greatest English physicist was superseded by a theory of a Swiss Jew living in Germany, on the basis of observations by English astronomers, made in Latin America and Africa.

committed a mathematical blunder. But it was Einstein who blundered, for Friedmann's mathematics was perfect. This was quickly called to Einstein's attention, and in May 1923 he published another one-paragraph comment:

> I have in an earlier note criticized the cited work [of Friedmann]. My objection rested—as Mr. Krutkoff in person and a letter from Mr. Friedmann convinced me—on a calculational error [of my own]. I am convinced that Mr. Friedmann's results are both correct and clarifying. They show that in addition to the static solutions to the [gravitational] field equations there are time varying solutions with a spatially symmetric structure.

By this time Einstein was the most famous physicist in the world. But mathematical solutions in physics are either correct or incorrect, and fame, reputation, and position have nothing to do with it. In this instance it was the obscure Russian physicist who was right and the world's most famous physicist who was in error. Although Einstein publicly retracted his criticism of Friedmann in a courteous and gracious manner, in private he still clung to his own model of a static universe.

In the mid 1920s there were three competing ideas about the universe: the static universe model of Einstein, the stationary universe model of de Sitter, and the expanding big-bang universe model of Friedmann. They were all mathematically consistent (although the static and stationary universe models necessitated the ugly cosmological term) and amenable to experimental verification. Observational astronomers were curiously silent on the issue, however. Some astronomers noted the problem, but rather than making a mad rush to their telescopes to see whether the universe was expanding, most astronomers seemed to consider the question one for theoretical physicists to argue about.

Even without the mathematical sophistication to follow the intricacies of Einstein's theory, one astronomer still sensed the importance of the issue. He might not have followed the unfolding saga in detail, for this astronomer had never even heard of Friedmann, who, shortly after his work on cosmology, had tragically died of typhoid fever at the age of thirty-seven. Although lost in the complexities of curved space, curvature tensors, and the new mathematics, this one astronomer somehow managed to glimpse the enormity of the problem. Perhaps it is not surprising that he seized the opportunity, for very few people have ever possessed the intuition and insight of Edwin Powell Hubble.

THE REALM OF THE NEBULAE

Knowledge of the universe is not reserved for any particular class. The stars seem to reach out and seduce all sorts of people. Even in as limited a geographical area as western Europe, cosmologists came from backgrounds as diverse as the Danish nobility, like Tycho Brahe; from German *pöbel* (rabble), like Johannes Kepler; from the Italian *classe media*, like Galileo Galilei; or from the country-gentleman class that is peculiarly English, like Isaac Newton. Rich, poor, noble, common—all were seduced by the heavens, and all contributed to our view of the universe.

In the beginning of the twentieth century the story shifted to another continent, where the New World led in the exploration of the new frontier of cosmology, extragalactic astronomy. From the great flat plains of the Midwest of the United States, seeming to step out from a stereotypical Norman Rockwell portrait of a middle-class, all-American family, a person emerged to lead astronomers into the realm of the nebulae: Edwin Hubble.

Hubble was born in November 1889 in Marshfield, a small town near the southern Missouri city of Springfield. In 1898 his family moved to the Chicago area, eventually settling in the suburban town of Wheaton in 1900, just a few blocks away from where a public school would one day be named in his honor. From 1900 to 1906 Hubble attended public school in Wheaton, graduating from high school in 1906 at the age of sixteen. Before graduating he left his mark at Wheaton High, not so much as a scholar (although he was a fine, but not the best, student) but as an athlete. Although two years younger than most in his class, he excelled in all aspects of track, particularly the shot put and high jump, winning the high jump at the state meet and breaking the state record. Pictures from his high school days show a tall, athletic, confident (perhaps cocky) young man, eager to face a future of limitless horizons.

After high school Hubble attended the University of Chicago on an academic scholarship. In his first two years at Chicago he excelled academically, winning the award as the best physics student in his sophomore year. His academic performance seemed to slip a bit during his last two undergraduate years, possibly because of outside distractions. He was a member of the track team, coached by the famous Amos Alonzo Stagg, and a member of one of the championship basketball teams produced in those years by the athletic powerhouse Chicago.[4] In a

[4] In the days before March Madness, the University of Chicago basketball teams of Hubble's final two years were probably the best college teams in the country.

Edwin Hubble as a student and as an athlete at the University of Chicago. Pictured on the right is the 1910 University of Chicago Big 10 championship basketball team. Hubble is the fourth player from the left, with his arm on the knee of the player behind him.

single basketball game Hubble once scored 12 points. That wouldn't qualify him for a shoe endorsement these days, but in 1910 college teams rarely scored as many as 30 points, and final scores of 17–15 were not uncommon. In his senior year he was also elected vice president of his graduating class. He even found time to get into trouble for tossing a few eggs at the divinity students. Hubble was liked by all and seemed to be one of those students everyone knew was destined for success in something.

Because of his combination of scholarship, leadership, and athletic ability, Nobel laureate Robert Andrews Millikan, for whom Hubble worked as a physics laboratory assistant, recommended him for a Rhodes scholarship, describing him as a "man of magnificent physique, admirable scholarship, and lovable character."

Hubble continued to excel at Oxford University, doing well in academic study as well as starring on the track team, swimming on the water polo team, and continuing his hobby as an amateur boxer (he later sparred with the heavyweight champion of France). At Queens College, Oxford, he studied law, probably because of the wishes of his father, who wanted his son to be a lawyer. He did very well as a Rhodes scholar, but the report of his supervisor at Oxford in the British style was somewhat more reserved than Millikan's: "Considerable ability. Manly. Did quite well here. I didn't care v[ery] much for his manner. Will get [a grade of] *A*." Perhaps the self-assured Yank rubbed his British supervisor the wrong way.

After three years at Oxford, Hubble returned to Louisville, Kentucky, where his family had moved while he was in England. But rather than practice law or pursue a career in politics like most Rhodes scholars trained in jurisprudence, he decided to do something noble with his life, and he became a high school teacher.

Just across the Ohio River from Hubble's Louisville home was the small town of New Albany, Indiana. In the high school there Hubble taught Spanish and physics, and coached the men's basketball team to the state playoffs (where they lost in the third round).[5]

Hubble seemed to like teaching but decided that astronomy was what mattered. At the end of the school year he wrote a former astronomy professor at the University of Chicago, Forrest R. Moulton, to ask

To
Edwin P. Hubble
Our beloved teacher of Spanish and Physics, who has been
A loyal friend to us in our senior year.
Ever willing to cheer and help us
Both in school and on
The field.
We
The class of 1914 lovingly dedicate this book

The 1914 graduating class of New Albany, Indiana, High School "lovingly" dedicated their yearbook, The Senior Blotter, *to their Spanish and physics teacher, and basketball coach.*

[5] Occasionally someone remarks that a problem with science education in the United States is that athletic coaches often teach science courses. I always think of Hubble when I hear that.

about graduate study at Chicago in astronomy. Moulton recommended him to Edwin B. Frost, director of the University's Yerkes Observatory:[6]

> Personally, [Hubble] is a man of the finest type. Physically he is a splendid specimen. In his work here, altogether, and especially in science, he showed exceptional ability. I feel sure you would find him just the sort of man you would wish to have.

It is curious today to read so many letters of recommendation for academic positions referring to a candidate's physique, lovable character, or manly behavior.

As a graduate student Hubble attended a meeting of the American Astronomical Society at Northwestern University in Evanston, Illinois, in August 1914. Here the astronomer Vesto M. Slipher reported observational evidence suggesting that several spiral nebulae seemed to be traveling away from our galaxy at incredible speeds, as much as six hundred miles *per second.* No one at the time knew what this meant (in fact few even believed Slipher), for it would be another fifteen years before Hubble would put this piece of the puzzle in its place and discover that it was a key in the discovery of the expansion of the universe. Although Hubble was just starting out in astronomy, perhaps Slipher's paper made an impression on him, for he decided to make the realm of the nebulae his own.

Hubble's 1917 Ph.D. thesis, "Photographic Investigations of Faint Nebulae," was not distinguished; it seemed rather rushed and sloppy in places.[7] In fact, the thesis was rushed, for the United States had entered World War I on April 6, and the All-American boy was anxious to serve his country. On May 12 Hubble defended his thesis, and three days later he reported for basic training as part of the American Expeditionary Force destined for France.

As a college graduate, Hubble was eligible for a commission if he could find five "worthy citizens" willing to recommend him. As a

[6] Edwin Frost served as director of Yerkes Observatory until 1932, despite the fact that he lost sight in one eye in 1916 and by 1921 was completely blind. While blind watchers of the sky are common, blind directors of astronomical observatories are rather rare.

[7] Hubble's adviser decided that the thesis was not worthy of publication in the most prestigious American astronomy journal, the *Astrophysical Journal,* but should only appear in the *Publications of the Yerkes Observatory.*

"manly" Rhodes scholar of "magnificent physique," "exceptional ability," and "lovable character," finding recommendations was never a problem for Hubble. Included among those who recommended him for a commission was George Ellery Hale, the former director of Yerkes Observatory before he left to build the (then) largest telescopes in the world at Mount Wilson Observatory outside of Pasadena, California. Hale had met Hubble during his many visits to Yerkes, and had his eye on the promising young astronomer for a future staff position at his observatory.

Tall, intelligent, athletic, handsome, self-assured, enthusiastic, and a natural leader, Hubble was quickly promoted to captain (eventually he would become a major) and placed in command of the 2d Batallion, 343rd Infantry Regiment, of the 86th Division of the United States Infantry. As with everything else he did, Hubble was successful as a commander; his only shortcoming seemed to be a frightening eagerness to lead his men into battle as quickly as possible.

It was early October 1918 before Hubble reached France, and to his dismay he "barely got under fire," as he put it, before the end of the war. After the end of hostilities Major Hubble was assigned to Cambridge University to oversee American army students in British universities. Cambridge was the home of the most famous astrophysicist in England, Professor Arthur Stanley Eddington. But Hubble would not interact much with Eddington at this time, because he was preparing for the Principe eclipse expedition whose results would rock the world and prove Einstein correct.

Hubble mustered out of the army in August 1919 at the Presidio in San Francisco, and headed directly to Mount Wilson, where Hale had offered the thirty-year-old Hubble a position of staff astronomer with the handsome salary of $1,500 per year and the opportunity to use the new hundred-inch telescope just starting operations.

Hubble first used the Mount Wilson telescopes on the night of October 25, 1919. The sixty-inch and hundred-inch telescopes were the perfect instruments to launch the study of extragalactic astronomy, and not since Galileo turned his first crude optic tube toward the sky had the right tool been in the right hands at the right time. From the beginning Hubble's interest was in nebulae, and in the words of another astronomer who witnessed Hubble's first night observing, "He was sure of himself—of what he wanted to do, and of how he wanted to do it."

Hubble truly explored the realm of the nebulae. In 1924 he proved beyond doubt that they are beyond the confines of our own galaxy. But

The sixty-inch (left) and hundred-inch (right) telescopes at Mount Wilson observatory. The sixty-inch went into service in 1908 and the hundred-inch in 1917.

that remarkable discovery set the stage for his even more important demonstration of the expansion of the universe. At the heart of this second discovery was Hubble's realization that there is a correlation between the distance to a galaxy and its "red shift."

THE RED SHIFT

The immutable eighth crystalline sphere of the heavens, which in the view of the ancient Greeks and early Renaissance astronomers carried the stars and planets on their appointed rounds, had become quite mutable by the end of the eighteenth century. Not only had the spheres of the planets been smashed in the sixteenth and seventeenth centuries, but by carefully monitoring the positions of stars, Edmund Halley proved in 1718 that stars are not locked into a single position in space, but they have motions of their own across the **celestial sphere**.

If stars are not locked in position and they move around on the celestial sphere, it would make sense that they are also free to move closer to us or farther away from us. But even an accurate determination of the position of an object in the sky cannot determine the velocity of the object toward us or away from us. This can be discovered only through a phenomenon known as the Doppler shift.

Occasionally basic research has a payoff for society, but sometimes it can take a few years, or even a century or so, before society reaps the benefit. For example, in 1842 an Austrian physicist and native of Salzburg, Johann Christian Doppler, made a wonderful discovery that would not have its real impact on society until the second half of the tweentieth century.

In 1842 Doppler proposed that sound emitted with one pitch from a moving object would be detected as a different pitch by a stationary observer. Today, anyone who has had the common experience of listening to a train passing by, a plane traveling overhead, or a car speeding past knows that the sound changes. But life wasn't so fast paced in 1842 when Dopper proposed his effect, so it was not part of everyday experience and people had to test it for themselves. Perhaps the most dramatic confirmation of the Doppler effect was by a Dutch meteorologist Christopher Heinrich Dietrich Buijs-Ballot.

Buijs-Ballot convinced his friend the Honorable L. J. A. van der Kun, the Dutch minister of the interior, to let him borrow a railroad and locomotive for a day, with the promise that he would give it back just as soon as he finished a simple scientific experiment. So outside of Utrecht one fine Saturday in June 1845, Buijs-Ballot assembled a team for an experiment to test Doppler's idea.

Buijs-Ballot's experiment was rather unusual because it involved not a group of scientists but an ensemble of musicians. The idea was pretty simple: Buijs-Ballot put a musician in an open car pulled by a locomotive through the Dutch countryside near Utrecht and instructed him to play a single note, one octave and one whole note above middle *A*, which has a frequency of about 1,000 Hertz (the theory of the Austrian Doppler can be tested by a single note, and does not require a composition as imaginative as those of another famous native of Salzburg). Meanwhile, as the train passed by, other musicians, acting as critics, would stand alongside the track and record the note they heard. If Doppler was correct, the track-side observers would hear a higher note as the train approached and a lower note as the train receded. Buijs-Ballot would then subtract the frequency of the note the musician played from the frequency of the note the critics heard, and compare it with the change in the frequency due to the motion of the train as predicted by Doppler, and either confirm or disprove the theory.

Excitement was in the air on June 3, 1845, as a crowd of curious onlookers gathered outside Utrecht. Buijs-Ballot carefully measured the temperature, barometric pressure, and the humidity (after all, he was a meteorologist) in order to calculate the sound speed. He then gave the signal: "Gentlemen, start your engine," and the train slowly gathered steam and chugged along the Dutch countryside past windmills and tulips, eventually reaching a top speed of thirty miles per hour (not bad for 1845). The trumpeter blew with all his might as the train sped past the observers. Buijs-Ballot ran to the music critics on the side of the tracks and eagerly inquired, "Well, what note did you hear?" In unison the observers answered, "What, did you say something?" After a few minutes the observers recovered their hearing, which had been lost in the clamor of the passing train, and informed Buijs-Ballot that they couldn't hear a damn thing over the noise of the train.

It isn't unusual that the first attempt at an experiment fails because of insufficient resources. Rather than give up, Buijs-Ballot stuck a few more musicians on the train and instructed them to blow harder. The musician's union was not so strong in those days, and it was possible to demand that trumpeters play until they were blue in the face while bouncing along the bumpy train tracks at thirty miles per hour, with sparks and soot flying about their heads. For the rest of that Saturday the train made another five passes, but the results were inconclusive. After convincing the government that he required use of the railroad for just one more day, Buijs-Ballot retired for the evening and allowed the musicians to catch their breath on Sunday before making another go of it on Monday.

For the Monday trials Buijs-Ballot had the train make six trips past three different observing stations at speeds between eleven and forty-five miles per hour. This time the experiment worked. The critics informed Buijs-Ballot that they heard higher frequency (shorter wavelength) notes as the train approached, and lower frequency (longer wavelength) notes as the train receded, with the frequency shift as large as an entire semitone, depending on the speed of the train. The musicians were the best that money could buy, and Buijs-Ballot was sure they were playing the correct note. He concluded that the reason the note sounded sharp when the train approached and flat when it receded was that the pitch, the frequency of the sound, was affected by the motion of the source as Doppler predicted. Furthermore, the faster the motion of the sound source, the larger was the shift in the detected frequency, in direct proportion to the velocity.

If the pitch of the detected note is proportional to the relative velocity between the source and observer, the velocity can be determined if the frequency of the detected note and the emitted note are known. For a velocity of thirty-five miles per hour, the shift would be an entire semitone. A reasonably good musician can tell the difference of about a twentieth of a semitone, or a shift in frequency of about 3 Hertz around *A*. This shift in frequency corresponds to a relative velocity as slow as five miles per hour between the musician and listener. So by determining the difference in pitch, the velocity can be determined quite accurately.

The Doppler effect works for light as well as sound; in fact, it applies to all wave phenomena, including parts of the **electromagnetic spectrum** not visible to us, like radar waves, X rays, radio waves, and so on.[8] The source of the radiation really doesn't have to originate with the moving object, it could have been reflected from the object. In fact, one might imagine pointing a source of electromagnetic radiation of a known frequency at an approaching or receding object, and detecting the reflected radiation from the moving object at a different frequency because of the Doppler effect.[9]

The next time you are driving on the highway and notice someone on the shoulder of the road pointing a source of electromagnetic radiation in your direction, measuring the frequency shift of the radiation reflected from the windshield of your car, comparing the reflected frequency to the frequency emitted by the radar gun, and determining your velocity, you can thank Christian Doppler for keeping our highways safe. When the person who pointed the source of electromagnetic radiation at you is standing outside your automobile asking you if you realized how fast you were driving, you can think of Doppler, Buijs-Ballot, and a train of trained musicians, and be thankful for the benefits of basic research.

The reason we are interested in the Doppler effect is that it can be used to determine if celestial sources are approaching toward us or receding from us. As far as we know there are no musicians in the sky, so we can't use sound waves to measure velocities, but we can use light waves. Luckily, radar guns aren't needed to bounce electromagnetic

[8] Curiously, most of Buijs-Ballot's paper was devoted to an incorrect argument that the Doppler effect did not apply to light waves.

[9] It doesn't matter if the source is moving or the detector is moving. The only important motion is the *relative* motion between the source and observer. Buijs-Ballot also tested this by stationing the musicians on the side of the tracks and placing the critics on the moving train.

waves off stars to measure their speeds, because in stellar spectra there are lines of known frequencies due to atomic transitions.

If the star emits light of a particular wavelength, the light will appear to have a higher frequency (shorter wavelength, that is, bluish) if it is approaching and a lower frequency (longer wavelength, that is, reddish) if it is receding, in exactly the same way Buijs-Ballot's trumpeters' note sounded sharp when approaching and flat when receding. So the velocity of celestial sources can be determined by comparing their spectra with the spectra of stationary sources.

By the beginning of the twentieth century astronomers were systematically measuring the velocity of stars by this method. They found that just as stars move across the sky, some stars are traveling directly toward us and some straight away from from us, with typical velocities of a few dozen miles per second. Eventually telescopes and spectroscopes became good enough to use the technique to measure the velocity of faint, diffuse spiral nebulae. One of the first to do this was Vesto M. Slipher.

Slipher was an astronomer at the Lowell Observatory, built by Percival Lowell in the San Francisco Mountains outside of Flagstaff, Arizona, to search for life on Mars. Ironically, perhaps the greatest discovery ever made at the observatory had nothing to do with life or canals on Mars or any other planet in our solar system, but with the determination of the velocities of objects in the universe far outside our solar system.

Lowell had instructed Slipher to build an instrument to measure the spectra of the light from the planets in order to learn about their rotation. If the planets are spinning, then there will be a small red shift of the light from the outer edge of the planet going away from us and a small blue shift at the outer edge of the planet rotating toward us. By measuring the Doppler shift of the spectral lines at the limbs, Slipher could determine the rotational velocity of the planets. Lowell then assigned Slipher to take spectra of the Andromeda nebula, because it was thought to be a solar system in its embryonic stages.

Slipher did not measure a rotational velocity for Andromeda, but found the remarkable result that the entire nebula was coming toward us (blue shift) with a velocity of nearly two hundred miles per second! He then found three other spiral nebulae with recessional velocities (red shifts) of over six hundred miles per second. These velocities were so much larger than the typical velocities of nearby stars in our galaxy that at first his results were not believed. Just as we always think that we couldn't possibly have been driving as fast as indicated by the Doppler effect radar and something must be wrong with the radar gun, astronomers didn't believe Slipher's high velocities for the nebulae. His report of these velocities at the 1914 American Astronomical Meeting in Evanston caused quite a stir; they were dismissed by many, but they obviously made a big impression on a young, beginning graduate student in the audience named Edwin Hubble.

Soon others confirmed Slipher's measurements. In 1918, in the muddy trenches of France on the opposite side of the western front from Edwin Hubble, the German astronomer Carl W. Wirtz took time out from the war to write a paper suggesting that all spiral nebulae recede from us with a typical velocity of about four hundred miles per second. It was becoming clear that something was very funny about the velocity of spiral nebulae.

Slipher pushed ahead with the program of measuring the recessional velocities of spiral nebulae. In 1921 he measured what was then the world-record recessional velocity for a spiral nebula of over eleven hundred miles per second, and started the tradition of announcing astronomical discoveries in the *New York Times.* In the newspaper article Slipher mentioned the possibility that the nebula was once in the vicinity of our galaxy, and had traveled with this large velocity out to a great distance from us during the history of the universe. This was perhaps the first mention in the popular press of a model that would later be called the big bang.

Slowly astronomers began to suspect that the recessional velocities of the nebulae were largest for the most distant nebulae. Any real quantitative test of this hypothesis was impossible, however, because distances to the nebulae could not be determined with any certainty, so people tried various indirect ways to judge the distances. In 1924 Wirtz made the (admittedly unrealistic) assumption that all nebulae were the same size. If this would be true, then the distance to a nebula would be proportional to its apparent diameter, which could be determined just by measuring how large the image appeared on a photographic plate. Doing this, Wirtz found that the recessional velocity increased with distance, and even went so far as to claim that it was confirmation of de Sitter's cosmology.

The Swedish astronomer Kunt Lundmark also attempted to test de Sitter's idea of a correlation of velocity with distance. In 1925 he published a paper in which he claimed evidence for the increase of velocity with distance, but complicated the issue by analyzing a more general case where the velocity is a constant, plus a term linear with distance, plus a term quadratic with distance. Sometimes greater generality only thickens the fog.

The best instruments for extragalactic astronomy at the time were the giant telescopes on Mount Wilson, where the measurement of recessional velocities started with the work of Milton L. Humason, a colleague and friend of Hubble. In contrast to the pedigree of Dr. Edwin Powell Hubble (B.S., M.S., Rhodes scholar, Ph.D.), Mr. Humason's education ended after four days of high school. The high school dropout took the best job he could find—a muleteer, ferrying material up the mountain for the building of the Mount Wilson Observatory. Humason seemed to like it around the observatory, and in 1917 he was hired there as a janitor. After a few years he rose to the rank of night assistant, with the job of helping astronomers with their observations. Although he knew nothing of mathematics and had no formal training in astronomy, he was one of those people who seem to have a magic touch with anything mechanical, and this ability kept the balky instruments functioning when the professional astronomers on the staff had given up in frustration. As he learned more and more, he was given greater responsibility, and eventually he was allowed to make his own observations. His quiet, patient, conscientious manner made him the perfect person for the demanding task of taking the spectra of faint nebulae.

Ten years after starting as a janitor, the high school dropout and former muleteer Humason published his first scientific paper on the recessional velocities of spiral nebulae. The data analysis and mathematical calculations were done by others on the staff, and the actual text of his paper was ghostwritten by Hubble. But this should not overshadow the remarkable fact that in just a few years Humason had made the transition from muleteer to astronomer. The stars had seduced yet another helpless victim. Yet again we see that the allure of the stars is democratic; from Danish nobleman to American muleteer, they seem to strike indiscriminately.

By 1929 Hubble and Humason had measured the spectral shift of forty-six nebulae, and Slipher the shift of thirty-nine (some of them were the same nebulae). The base of data was increasing, but it was not easy to make sense of it.

Most astronomers realized that the crucial question was whether the recessional velocity was correlated with distance to the nebula. The real problem in untangling this issue traces back to the old problem of knowing distances to the nebulae, which required yet another step on the journey that started with the hole in the ground in Aswan.

Hubble could determine distances to only a few nearby galaxies using the technique he pioneered in 1924 of measuring the period of Cepheid variable stars, because even the bright Cepheid stars were too dim to resolve in distant galaxies. However, Hubble reasoned that if he knew the distances to the few nearby galaxies using the Cepheids, he could determine the intrinsic **luminosity** of the brightest stars in them by measuring their intensities and using the **inverse-square** law. When he did this, Hubble noticed that the very brightest stars in the galaxies with known distances all had about the same intrinsic luminosity. If he then assumed that the brightest stars in more distant galaxies all had the same intrinsic luminosity as the brightest stars in the nearby galaxies, then he could determine the distance to the far galaxies, again by using the simple inverse-square law.

Thus Hubble was able to find "reliable" distances to galaxies far enough away for the first real test of the de Sitter hypothesis. By 1929, Hubble had enough data (distances to twenty-four nebulae) to publish a paper, "A Relation Between Distance and Radial Velocity among Extra-Galactic Nebulae," in which he proposed a linear relationship for the velocity with distance. The closing paragraph of Hubble's paper contained a sentence that signaled that the fog shrouding the expansion of the universe had finally lifted:

> The outstanding feature, however, is the possibility that the velocity-distance relation may represent the de Sitter effect, and hence that numerical data may be introduced into discussions of the general curvature of space.

While others danced around the issue, Hubble, "sure of himself—of what he wanted to do and how he wanted to do it," had the decisive word on the subject, much as he had the final word on the island-universe hypothesis. Hubble had made another remarkable discovery, for not only were the spiral nebulae distant galaxies as he had shown in 1924, but now he proved that they were expanding away from us. Not only was the universe larger than previously believed, but it was growing larger every moment.

Once Hubble's paper appeared, most physicists realized that expansion and Einstein's theory are naturally compatible. In 1931 Einstein himself recommended dropping the cosmological term from the

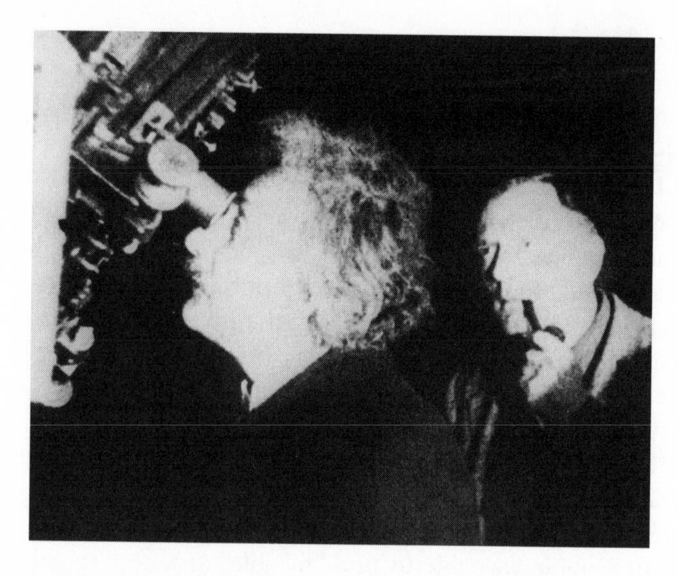

Hubble and Einstein at Mount Wilson during Einstein's visit of 1930. No one knows what Hubble seems to be whispering in Einstein's ear as the great theoretical physicist peers through the telescope. One legend has it that Hubble is saying, "Herr Professor, we should open the dome first."

gravitational equations. The big-bang picture finally emerged from the fog.

STRAIGHT AND TRUE ENOUGH

Why does Hubble get the credit? Why not Slipher, Lundmark, Wirtz, or any number of other astronomers who were so close to discovering the expansion of the universe? What about Friedmann, de Sitter, or Weyl, all of whom to some extent predicted the expansion and the red shift? Is it fair to say that Hubble discovered the expansion of the universe? What did Hubble do that the others did not?

In my opinion, Hubble justifiably deserves credit for the discovery of the expansion of the universe. One reason is that Hubble established distances to the galaxies, and only if the distances are known can one say for sure that the red shift increases in proportion to the distance. Another reason is that his paper contains something lacking in the others: a clear and concise statement of the result, and demonstration of a strong conviction that it is true. The message is stated in the title of the paper and is made clear throughout the paper with no ifs, ands, or buts. While others cloaked the possibility of the expanding universe in threads of complexity, or hid behind the tentativeness of "uncertain data," Hubble stood up with an air of confidence and said, "This is the way it is." If someone really knows what he is doing, and is confident in the result, he doesn't bury the result in a footnote or hide it in an inconspicuous paper.

✦ ✦ ✦

Shortly before Andrei Sakharov's death, I attended a conference honoring his scientific accomplishments. Although notoriety in his later years was due to his brave political stance for human rights, throughout his lifetime Sakharov made many important contributions to physics and cosmology.

Among the many physicists at the meeting recounting Sakharov's discoveries was one who made the surprising claim that in a footnote in an otherwise unknown paper, Sakharov had proposed a theory that a decade later proved to be the correct theory of the forces responsible for binding quarks inside neutrons and protons.

There is really no need to look through Sakharov's footnotes to find evidence of his remarkable insight as a physicist. Furthermore, I am convinced that he must not have understood the implications of what he said in the footnote, or perhaps he didn't really believe it, because it is inconceivable to me that anyone would make a discovery of that magnitude and hide it in a footnote.

✦ ✦ ✦

Hubble didn't hide his result in a footnote, but stuck his neck out. Indeed, to claim discovery of the expansion of the universe was a risky statement at the time—the data were not overwhelming. If we plot the recessional velocity versus distance as Hubble did, it is not totally obvious that it is a straight line. In fact, it is not obvious at all.

Anyone who has taught a high school or introductory college physics course knows that no matter what the data look like, a student invariably will draw a straight line through the points (even if the answer isn't supposed to be a straight line).[10] Looking at the data that

[10] For some reason, students who draw the line first and then plot the data always seem to find better agreement between the line and the data.

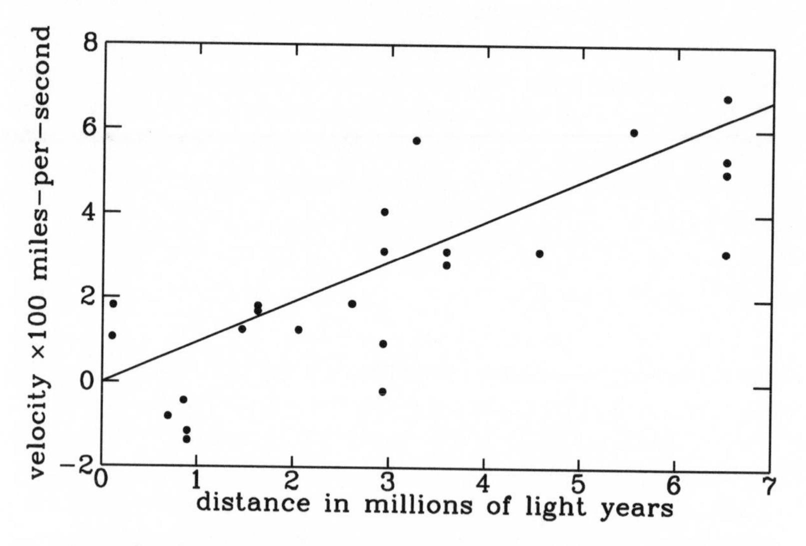

A graphical representation of the data on the distances (in millions of light-years) and velocities (in hundreds of miles per second) of the twenty-four nebulae Hubble used in his 1929 paper on the expansion of the universe. The line is Hubble's proposed fit to the data showing a linear increase of velocity with distance.

Hubble had to work with, we can see that it is not clear that a straight line should be drawn through the points. Part of the reason for the poor agreement is that there is a secondary effect that spreads the points around. Just as Galileo had to overlook the fact that a feather will not drop at the same rate as a bowling ball because of the secondary effect of air friction, part of Hubble's insight was to overlook the fact that his expansion law isn't exact, but it is an approximation (and not a very good one for nearby galaxies), because of the secondary effect of the motions of galaxies due to the gravitational effect of neighboring galaxies.

In fact, some of the nearby galaxies, like Andromeda, are actually approaching us. We now know that in addition to the recessional velocity predicted from Hubble's law, all galaxies have small random velocities that either add to or subtract from the Hubble velocity. When we measure the red shift of a galaxy, we measure a part due to the expansion of the universe, given by Hubble's law, and a part contributed by the random motions. The magnitude of the random velocities is about the same for every galaxy, while the recessional velocity according to Hubble's law increases with distance. So the random velocities can overwhelm the Hubble velocity for nearby galaxies, but they are an insignificant fraction of the total velocity of very distant galaxies. The farther we look, the more accurate Hubble's law becomes.[11] But the random velocities clouded the issue for the nearby galaxies Hubble knew about in 1929. In fact, when astronomers display a "modern" Hubble diagram, they don't even bother putting Hubble's original data on the graph. Now we know that if we extend the graph drawn by Hubble by looking at galaxies beyond his reach at the time of his first paper, the secondary effects are unimportant. But Hubble couldn't be sure.

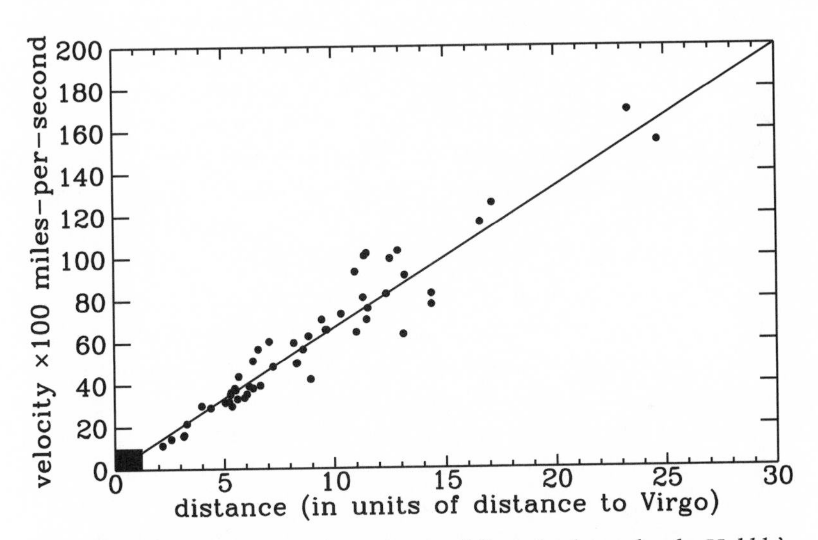

A "modern" Hubble diagram constructed from the data taken by Hubble's successor, Alan Sandage. The data on this graph start just about where Hubble's original data ended. Since the distances to the galaxies are uncertain, they are given in units of the distance to the Virgo clusters of galaxies, believed to be thirty to sixty million light-years distant. The spread in the data where there are departures from a straight line fit is thought to be caused by uncertainties in determining the distances to the galaxies. The region covered by Hubble's original data is shown by the box in the lower left.

[11] Unfortunately, the farther we look, the more difficult it becomes to determine the distance to the galaxy. Also complicating matters is the fact that the light we see from galaxies far away was emitted when galaxies were young, and young galaxies may have a different brightness from galaxies today.

It required no small amount of courage for Hubble to state his law. At the front, in the fog of physics on the raw edge of ignorance, the answers are shrouded in mist and are not as clear as they later become.

WHY A BANG?

The discovery that galaxies are receding from us was the first observational evidence that the universe is expanding. Furthermore, Hubble's law, that the expansion velocity increases in a linear fashion with distance, seems to point to an initial explosion.

We can easily see how Hubble's law implies a big bang. Perhaps dealing with distant galaxies and velocities of tens of thousands of miles per second is intimidating, so consider the more down-to-earth, familiar situation of trains. Anyone who enjoyed the old high school math problems involving a train leaving Des Moines traveling west at eighty miles per hour and a train leaving Sioux City traveling east at sixty miles per hour will have no trouble following the argument.

Imagine living in a central transportation hub such as Chicago. Suppose from our vantage point in Chicago we receive information that a nonstop train sixty miles outside of the city is traveling away from Chicago at sixty miles per hour. Then we might guess that it left the Chicago station about an hour ago, since sixty miles per hour times one hour is sixty miles. At the same time we are also informed that another train thirty miles away is traveling thirty miles per hour, so we would deduce that it also left the Chicago station one hour ago (since thirty miles per hour times one hour is thirty miles). If a third train fifteen miles away is traveling fifteen miles per hour, it also must have left the station one hour ago. If all the trains left the station at the same time but at different speeds, then the faster trains will be farther away, with the distance in direct proportion to the velocity (Hubble's law). If the speed of the trains traveling away from Chicago increased linearly with distance from Chicago, they must have all left at the same time.

If the velocity-distance relation were not linear, then we would deduce that all the trains did not leave the station at the same time. For instance, if the train sixty miles away was traveling thirty miles per hour and the train thirty miles away was also traveling thirty miles per hour, then the more distant train must have left the station earlier.

The same holds true for galaxies. Let's go out to any handy telescope with a mirror larger than, say, six feet in diameter and locate a very distant galaxy. Then pass the light from the telescope through a spectrograph and measure the shift of the lines in its spectrum to determine its recessional velocity from the Doppler effect. Let's say we determine its recessional velocity to be twenty thousand miles per second. Now let us find a second galaxy and do the same thing. Suppose we find that this one has a recessional velocity of ten thousand miles per second. If Hubble's law is correct, then this second galaxy with half the velocity of the first one is exactly half as far away.

We can do the same thing for more and more galaxies: if the velocity increases in proportion to the distance, then there was a single time in the past when all the galaxies were in the station together, and some galaxies are farther away today because they have a larger recessional velocity. If we assume that the galaxies have always moved with the same velocity, we can estimate when all the galaxies were together, just as we can calculate when all the trains left the station. With the recessional velocities we see now, this time would be ten to twenty billion years ago.

An implication of Hubble's distance-velocity relation is that the universe started from a dense state ten to twenty billion years ago, in a big bang. This model of an expanding universe superseded Newton's and Einstein's view of an unchanging universe. The twentieth-century view of the universe is dynamic—the universe is different today from yesterday, and it will be imperceptibly different tomorrow. Although we have always sensed change in nature, it typically has been a cyclic variation: night turns to day, and to night again; winter turns to spring, spring to summer, summer to autumn, and eventually back to spring again; planets move around, but sooner or later they return to the same spots in the sky. But there is no apparent regularity in the big bang. In some sense every instant is unique in an expanding universe, for our universe will never again look exactly the same as it does now. Aristotle believed that changes we see on Earth were only an artifact of our presence, that the universe by and of itself was unchanging and only our corrupting influence altered its perfection. But now we know that in some sense humans are no different from the universe: the only constant is change.

✦ ✦ ✦

But where is the station from which the galaxies departed? The train analogy made it seem that if we see every train traveling away from Chicago, it must be the center of the train world. But just because all the galaxies appear to be receding from us does not make the Milky Way the central station of the universe.

One of the hardest concepts in modern cosmology to understand is the idea that the big bang is an explosion without a center and without an edge. If the big-bang model is correct, then *any* observer on *any* galaxy *anywhere* in the universe would see galaxies receding. The key to understanding this seeming paradox is to repeat three times per day *"The expansion of the universe is an expansion of space, not an expansion of galaxies into space."*

To see how this might work, let's ignore for the moment that the surface of Earth is curved, and imagine that Earth is an infinite, flat surface (a good approximation around Chicago). Replace the thin sheet of rock, soil, and water on the infinite, flat surface of Earth with a sheet of rubber. Next, shut off the engines and stop all the trains dead in their tracks, just like what happens when you ride the trains in Italy and there is a spontaneous nationwide strike. Rather than have the trains move across a fixed surface, arrange for the surface beneath the trains to move and carry the trains along by stretching the sheet in all directions.

The distance from Chicago to every train will grow as the sheet is stretched, and to an observer sitting back at the Chicago station it will appear as if all the trains are receding, with the velocity increasing as the distance. But if we put a lot of trains across the country and stretch the sheet in all directions, any observer in any city, even in Indianapolis, would see the trains receding. The recessional velocity is due not to the motion of trains across Earth but to Earth's being stretched. *"The expansion of the universe is an expansion of space, not an expansion of galaxies into space."*

Imagine the infinite space of the universe studded with galaxies. Now stretch the space in all directions. Every galaxy would see every other galaxy receding, with the velocity of recession increasing with distance. Such a universe has no center and no boundary. *"The expansion of the universe is an expansion of space, not an expansion of galaxies into space."*

That an infinite space can have an explosion without a center or boundary is easy to see, but an explosion in a finite space without a boundary or an edge is not so simple to visualize. In fact, it is not easy to imagine a finite space that doesn't have a center or an edge even without the complication of expansion. For instance, Galileo thought that if our universe was finite, it must have a center.

In Galileo's *Dialog*, before refuting Simplicio's argument that Earth is at the center of the universe, Galileo's spokesman Salviati raises the prescient question of whether the universe has a center at all:

> **Salviati:** I might very reasonably dispute whether there is in nature such a center [of the universe], seeing that neither you nor anyone else has so far proved whether the universe is finite and has a shape, or whether it is infinite and unbounded. Still conceding to you for the moment that it is finite and of bounded spherical shape, and therefore has a center, it remains to be seen how credible it is that the Earth rather than some other body is to be found at that center.
>
> **Simplicio:** Aristotle gives a hundred proofs that the universe is finite, bounded, and spherical.
>
> **Salviati:** Which are all reduced to one, and that to none at all. For if I deny him his assumption that the universe is moveable all his proofs fall to the ground, since he proves it to be finite and bounded only if the universe is moveable. But in order not to multiply our disputes, I shall concede to you for the time being that the universe is finite, spherical, and has a center.

That Galileo would even raise the issue is remarkable, so we shouldn't quibble that his mathematical reasoning is flawed. Galileo was right that an *infinite* space has no center, but Galileo didn't know that it is possible to have a *finite* space without a boundary or a center. Galileo didn't realize that the universe could be finite and spherical but *without* a center because it would be another 283 years before Einstein would discover that space could be curved.

Curved space is hard to think about. Most people with limited imagination (like me) are comfortable thinking only about flat space. Whether or not they admit it, most people who have to deal with curved space in their job (not a really large sample of the population) usually think about it by surrounding the curved space with a flat space. For instance, imagine the simplest curved space I can think of, a circle. The circle is a one-dimensional space because any point on the circle can be described in terms of a single number, say, an angle. Everyone knows that the circle is a curved space—if you don't believe it, try drawing one with a straightedge. Although the curved space of a circle has a finite length, it is unbounded. If you walk around in circles, you might come back to your starting point, but you never leave the circle. If you are asked to visualize a circle, you probably close your eyes and picture a circle neatly drawn on a piece of paper. Although the circle is a one-dimensional curved space, we usually visualize it on a flat two-dimensional space, in this case the sheet of paper.

An example of a two-dimensional curved space that is finite but unbounded is the surface of Earth. We know that it is a two-dimensional space since any point on the surface of Earth can be specified in terms of two numbers, say, longitude and latitude. The surface of Earth is also a curved space. One way we can tell that it is curved is by picking any three points on its surface and connecting them by the shortest paths across the surface. Any triangle constructed in this way will have the sum of the interior angles greater than 180 degrees. (Most things you were taught in high school geometry are true only for flat spaces.) Although it is a finite space with finite surface area, the surface of Earth is unbounded. This is why no one has ever managed to sail off the edge of Earth. There is also no center and no edge to the *surface* of Earth.

When we visualize the surface of Earth we first imagine a three-dimensional flat space and then stick the surface in it. But the third dimension running into Earth and out into space is not a part of the space. Again, it is easier to think about a curved space by pasting it in a flat space of higher dimension, in this case a two-dimensional curved space in a three-dimensional flat space.

Return to the trains on the rubber sheet, but now, rather than an infinite, flat surface, let's keep the rubber sheet in the shape of the surface of Earth, so our model for Earth looks like a giant balloon. Again, stop all the trains in a worldwide rail strike, but now let the balloon expand. Then on the surface of Earth every observer will see every train seem to speed away although the trains are at rest, because distances on the surface of Earth are increasing as the balloon is blown up. Every observer will see a linear relationship of distance and velocity, but there is no center to the expansion and no edge.[12] The reason the balloon-train universe seems to expand is that space is being created as the balloon expands. "*The expansion of the universe is an expansion of space, not an expansion of galaxies into space.*"

Envisioning the two-dimensional curved space was easy, because it was convenient to embed it in the familiar three-dimensional flat

[12] Recall that the interior and exterior of the balloon are not part of the space.

space of our everyday existence. But now for a model of the universe it is necessary to imagine a three-dimensional curved space. To do this, start by imagining a *four-dimensional* flat space (it helps to close your eyes), and place in it the three-dimensional curved space of the universe.[13] The three-dimensional curved space of the universe will have no center and no edge (or the center and the outside of the three-dimensional space is in the fourth spatial dimension, which isn't part of the universe).

So there you have it—an example of a three-dimensional space that has no center, has no edge, and has a finite volume.[14]

We don't know if our universe is infinite in space and flat, or finite and curved (there is a third possibility—infinite and curved—but to think about that three-dimensional curved space requires visualization of a *five*-dimensional flat space). Although we do not know if it is curved, we can determine that we would have to travel at least ten thousand million light-years before noticing any signs of curvature.

There are still many questions about the expansion of the universe: Will it expand forever or ultimately recollapse? Is the universe curved or flat? Did the expansion commence ten billion years ago or twenty billion years ago? Although the fog shrouding the expansion of the universe has been lifted, there are still things we cannot discern.

The discovery of the expansion of the universe settled some issues but raised new questions. Some of the questions would never have been asked without the big-bang model providing a *schema*, or framework for examining the universe and seeking a deeper understanding. We don't know whether all these new questions are answerable within the big-bang model itself, or if we will be forced to find a larger, encompassing *schema*. In the meantime, as a cosmologist I can enjoy the questions we now have, for what wonderful questions they are.

[13] Warning: don't use time as the fourth spatial dimension.

[14] If you think the volume of the three-dimensional spherical space constructed in this way is $4\pi/3 \times R^3$, back up two pages and start again. The correct volume is $2\pi^2 \times R^3$. The space with the volume $4\pi/3 \times R^3$ is a three-dimensional space with a center and an edge.

NINE

A Matter of Degrees

Just one year after the bang, in the formless, fiery mist that existed long before the formation of planets, stars, galaxies, or even atoms, a quantum of electromagnetic radiation—a photon—passed close to an electron and was deflected by the electromagnetic force. The deflection was violent enough to create a new light quantum, which in turn scattered off other electrons in the primordial soup, sometimes picking up a little energy, sometimes losing a bit.

The newborn bundle of electromagnetic energy would scatter with electrons more than three hundred million times in the next three hundred thousand years, until finally the temperature of the universe dropped below 6,000° Fahrenheit, cool enough for atoms to form. After almost all of the electrons were swept into electrically neutral atoms, the photon was no longer chained by its interactions with the charged electrons, and at long last it was finally free to roam unimpeded across the universe.

Although by the time this photon found its freedom it was just one of 547 thousand million light quanta in every cubic centimeter of the universe traveling about in every conceivable direction, it was destined for a special place at the end of a fourteen-billion-year-long trek across the universe.

About a billion years into the photon's long trek, if it had been able to look about, it might have noticed the first stars and galaxies lighting up the blackness of space. Another seven billion years after that, when the photon was still billions of light years away from us, Earth started its slow condensation from the debris left behind in the formation of the solar system.

As the photon continued its uninterrupted voyage at the velocity of light across the silent depths of intergalactic space, on our planet continents began to take shape and slowly drift across the blue oceans. About eleven billion years after the photon began its long journey, the first microscopic organisms developed on Earth. When the first species of *homo erectus* appeared on the dusty plains of Asia about two million years ago, the photon was among our local group of galaxies, passing nearby the Andromeda galaxy.

By the time the photon entered into the region of the universe occupied by our galaxy, our ancestors were emerging from the deserts, forests, and jungles and forming the social organizations we call civilization. Traveling to its ultimate destiny, our intrepid photon passed by stars and traveled through interstellar clouds while the great civilizations of our planet grew, flourished, and disappeared. When Khufu erected his great pyramid at Giza, the photon was still thirty million billion miles away, and our Sun would have appeared to the photon as just another point of light among the multitude of stars in the Milky Way.

As time passed the photon drew ever closer to us. For every year that went by, the photon traveled six million million miles across the vast reaches of space toward its final destination. As the world mourned the assassination of John Kennedy at the end of November 1963, the photon was a mere three thousand billion miles away, and headed directly toward the brightest point of light it could see, our Sun.

Finally, at 11:15 A.M. EDT on May 20, 1964, after traveling unimpeded across nearly twenty billion light-years of space in the expanding universe, after witnessing the birth and death of galaxies, stars, and planets, the photon's silent sojourn ended when, while passing through our solar system, it happened to bump into *New Jersey!*

Of course, photons have no feelings, so it is senseless to wonder how the photon "felt" to end its existence in such a manner. But if we put ourselves in the place of this particular photon, we might think that it found solace in the fact that it didn't hit just any old spot in New

Robert Wilson (left) and Arno Penzias (right) in front of the giant twenty-foot antenna of Bell Telephone Laboratories' Crawford Hill Laboratory in Holmdel, New Jersey. The equipment trailer attached to the left of the antenna is where the photons were directed to be detected after being collected in the antenna.

Jersey, but happened to fall right into the most sensitive antenna that existed anywhere in the world at the time. If we can imagine that our anthropomorphic photon felt some degree of misery to have traveled 20 billion light-years across the universe only to crash into New Jersey, we would also have to surmise that it probably loved the fact that it had a lot of company, for in every second, on every square centimeter of the antenna, some three thousand billion fellow photons met an identical fate. Indeed, every square centimeter of Earth is hit by a similar number of ancient photons every second.

Almost all of the countless multitude of such photons that hit Earth every second have no effect on humans, but the photons that fell into the antenna that day in May had a very real effect on two humans sitting in a control room just a few feet from the throat of the antenna: they caused a great deal of consternation.

Since we have tracked the photon from its birth a year after the bang until the time it entered the antenna in Holmdel, New Jersey, we might as well tag along with it for a few tenths of a microsecond more.

Along with billions and billions of its companions that happened to fall on the 391-square-foot aperture of the horn-shaped antenna at the same time, the photon was directed to a six-inch-square opening at the throat of the antenna. Then the photon passed out of the antenna, through joints, couplers, switches, and masers that constitute the electrical "plumbing" of a modern radio telescope, finally being converted into an electrical signal. The electrical signal was amplified, modulated, shaped, and filtered, eventually appearing as a displacement of a pen on a chart recorder in the control room adjacent to the antenna. In the room, two young radio astronomers, Arno Penzias and Robert Wilson, looked at the tracings of ancient photons on the chart recorder, but didn't yell "eureka" at the moment of the discovery that would eventually merit them a Nobel Prize. Instead, they scratched their heads in bewilderment. Although their discovery of the background of ancient photons was one of the two great cosmological discoveries of the twentieth century (the other was Hubble's discovery of the expansion), at first the discovery was thought to be a nuisance.

A view of some of the plumbing in the inside of the equipment trailer. After entering the antenna, the ancient photon traveled through this maze of guides, switches, couplers, and so on, before appearing as a tracing on a chart recorder.

TO SEE THE LIGHT

When the ancient photons tumbled into the antenna in New Jersey, Penzias and Wilson stumbled on their discovery. Or at least that's the way many people have told the story. But before one attributes the discovery to lady luck, it should be appreciated that the most difficult thing to discover is something for which you are not searching, and Penzias and Wilson were not looking for the ancient photons. They had planned a routine measurement of the intensity of the radiation coming from the remnant of a supernova explosion, a mere ten thousand light-years from us, and were not expecting photons originating from near the edge of the observable universe.

But perhaps they should have been! The existence of the background photons had been predicted more than fifteen years previous to their discovery, and the Crawford Hill antenna was the ideal instrument to find them. In fact, the year before the discovery of Penzias and Wilson, two Russian astrophysicists, Igor Novikov and Andrei Doroshkevich, had published a paper in a Soviet physics journal stating that something must be wrong with the big-bang theory; otherwise the background photons would have been detected by the very Crawford Hill antenna being used by Penzias and Wilson.

If Penzias and Wilson were remiss in not being aware of the small body of literature discussing the possibility of ancient background photons, they certainly had a lot of company, for several other people had seen evidence for the ancient photons over the years, but ignored it. *(The most difficult thing to discover is something for which you are not searching.)* There seems to have been a complete disconnect between the theorists who predicted the existence of the background of ancient photons and astronomers who could have found them. Just as Penzias and Wilson were unaware of the theoretical papers predicting the ancient photons, for the most part the theorists making the prediction did not fully appreciate that instruments were available that were capable of detecting the radiation.

Even more curious is that while Penzias and Wilson were trying to figure out what was hitting their antenna, while Novikov and Doroshkevich were writing that the background photons couldn't be there because the antenna had not seen them, and while theorists who had predicted the existence of ancient photons were unaware that the technology to detect the photons had been in place for at least a decade, just down the road from Penzias and Wilson in Princeton, New Jersey, a group combining both theorists and observers were putting the pieces together on their own. But following the local tradition, the Princeton group did not assemble the puzzle by putting together the pieces others had left on the table; they started by stamping out their own puzzle pieces.

The Princeton collaboration dedicated to looking for the ancient photons was led by Robert Dicke, whose style was to do everything "in house." Jim Peebles, the theoretical arm of the group, had set about to rediscover on his own the theoretical arguments for the existence of a detectable background of photons, while Dicke, along with Peter Roll and David Wilkinson, was busy constructing a small antenna to receive the ancient photons that they planned to install on the roof of the physics building. Largely unaware of the literature, Peebles completed the calculations and the experimentalists worked in the basement to fashion an instrument capable of detecting the radiation, while just sixty miles beyond the ivy-covered walls of Princeton University, the perfect tool for the detection, the giant Crawford Hill antenna, had already detected the background radiation, without anyone realizing the significance.

Ironically, the fact that the Crawford Hill antenna should easily detect the background photons escaped the attention of the theorists who originally made the prediction, the Princeton group, and Penzias and Wilson, but not Novikov and Doroshkevich several thousand miles away in Moscow. For reasons probably unrelated to cosmology, the technical capabilities of Bell Laboratories were well documented in Moscow.

As in any human endeavor, there are complicated reasons why a prediction as fundamental as the existence of a background of ancient photons languished in dusty journals as people stumbled over the evidence while blindly watching the sky. But a large part of the problem was the fact that most scientists at the time just did not take the big-bang model, nor its proponents, as seriously as they should have.

In hindsight it seems remarkable that physicists and astronomers did not rush to embrace the big-bang model after Hubble's momentous 1929 discovery of the expansion of the universe. Although some quite famous physicists continued to dabble in the subject (Einstein, de Sitter, Weyl, and Eddington, to name a few), none of them would have identified himself as a cosmologist. Eddington was an astrophysicist, and the others were interested in cosmology as an arena for the application of the general theory of relativity. Other relativists in the United States had a keen interest in cosmology: Richard Chace Tolman at Caltech was close to Hubble and his extragalactic observational program, and on the

East Coast, Howard Percy Robertson at Princeton University made notable contributions. However, most leading physicists of the time following Hubble's discovery were quite caught up in the development of quantum mechanics and its application to atomic and nuclear physics, and just didn't have time to pay much attention to cosmology. Today an international meeting on cosmology can draw as many as one thousand scientists from around the world, but a meeting of cosmologists in the 1930s or 1940s could have been held in someone's living room.

There were other reasons that the fuse lit by Hubble's discovery burned for nearly forty years before the explosion of interest in cosmology we see today. A primary reason is that there seemed to be a fundamental contradiction in the model because it predicted that the universe is younger than Earth!

One by-product of Hubble's observational program of measuring the dependence of the red shifts of galaxies on the distances to the galaxies is an estimate for the time since the start of the big bang. It is easy to see how knowing the distance to a galaxy and its velocity provides an estimate for the age of the universe. The equation *time equals distance divided by velocity* can be solved using the measured distance to the galaxy and the measured recessional velocity to give the unknown time. Of course, the age obtained in this manner is just an estimate, because according to the big-bang theory the recessional velocity decreases as the expansion of the universe is slowed by the gravitational attraction of all the matter in the universe.

To estimate the age of the universe, one need know only the distance and recessional velocity of just one galaxy. For example, as part of his discovery paper Hubble determined that one galaxy he thought to be 6.5 million light-years away (about forty billion billion miles) had a recessional velocity of 625 miles per second (twenty billion miles per year). Using the *time equals distance divided by velocity* relation, it would have taken some two billion years for the galaxy to get there. Thus, if Hubble's law and Hubble's data are correct, the universe began some two billion years ago.[1]

Although two billion years might seem like a long time, by 1929 physicists such as Lord Rutherford calculated the age of Earth to be

[1] The same estimate for the age would have been obtained using any galaxy. If the recessional velocity had been, say, twice as large, then according to Hubble's law the galaxy would be twice as far away, and in dividing the distance by the velocity to find the time the factors of two would cancel.

about 3.4 billion years based on the relative abundances of radioactive elements in Earth's crust (the modern value for the age of Earth is 4.6 billion years). Thus the "Hubble age" for the universe was about half the age of Earth! No cosmology that predicts Earth to be twice as old as the universe gains many followers.

This seeming contradiction of the big-bang model was resolved in 1952 when Hubble's colleague Walter Baade at Mount Wilson discovered that there were two types of Cepheid variable stars, with different **period-luminosity** relations. When Hubble measured the period of Cepheid variable stars, he used Henrietta Leavitt's original relation to deduce the star's intrinsic luminosity from its period, and then he used the **inverse-square** law to calculate how far away the star had to be to result in its observed brightness. But Baade discovered that the Cepheid stars Hubble saw in nearby galaxies were of a different type than Leavitt used, and they were brighter than Hubble assumed. If they were brighter, they had to be farther away to appear so faint on the photographic plates. Thus the extragalactic distance scale was stretched by a factor of two.[2] If the galaxies were twice as far away as previously assumed, then the estimate for the age of the universe had to be revised upward by a factor of two. This solved the dilemma that the universe appeared to be younger than Earth.[3]

Although by the mid-1950s there was no longer an "age crisis" in the big-bang model, it still did not gain widespread acceptance by the scientific community. The general feeling was that in the balance between speculation and fact that characterizes any scientific discipline, cosmology was almost purely speculation with few testable predictions. One might argue whether this was necessarily the case, but it was a self-fulfilling prophecy: very few people worked in the subject because it was felt there were no hard predictions to be made, which meant that very few predictions were made because very few people worked in the subject. So the whole cosmology business was thought to be one that a

[2] We truly must climb a long ladder to learn the distance. The method of determining the distance to a galaxy depends on being able to determine the distance to stars in our galaxy, which in turn requires a knowledge of parallax, for which we have to know the Earth–Sun distance, and so on. The distance ladder is fragile in the sense that any readjustment of a lower rung affects the distances to everything on higher steps.

[3] In fact, the extragalactic distance scale was further revised upward in the 1950s to a value between five and ten times larger than Hubble's original calculations. The distance to all but the very closest galaxies is still in dispute today.

reputable physicist should stay away from, and the people who were involved with it were considered somewhat eccentric.

Well, there might have been some truth to the last allegation, for the two main champions of the big-bang theory in the 1930s, 1940s, and 1950s—Georgy Antonovich Gamow and Abbé Georges Henri Lemaître—were somewhat unusual.

George Gamow was born on March 4, 1904, in the Ukrainian city of Odessa. After graduating from Novorossia University in Odessa, he attended the University of Leningrad for graduate studies in physics. From early childhood he had a fascination with Einstein's theory of relativity, so he was an eager student in the course "Mathematical Foundations of the Theory of Relativity" taught by Professor Aleksander Friedmann. No doubt Gamow learned of the theory of the expanding universe in the course of lectures from the very person who first proposed the idea in its modern form. Gamow started work toward a graduate degree under Friedmann, but the unfortunate death of Friedmann in 1927 put an end to his plan. Gamow eventually finished his studies in 1929 with a thesis on a subject he described as "extremely dull." His graduation was delayed not by Friedmann's death but because of the Commissariat of Education's addition of two new required courses for physicists: "The History of World Revolution" and "Dialectical Materialism."[4]

Georgy Antonovich Gamow (left), and Abbé Georges Henri Lemaître (right).

[4] Although he passed both examinations, he claimed he never knew the meaning of the term "dialectical materialism."

Gamow then worked abroad at the University of Copenhagen and the University of Cambridge before returning to Leningrad as a professor of physics. While abroad, Gamow toured the countryside of western Europe on an enormous BSA motorcycle. He must have made quite an impression on the reserved British scientists when he roared up on his BSA to the front door of the Cavendish Laboratories in quiet Cambridge. On or off his motorcycle, Gamow seemed to make a lot of noise wherever he went.

When Gamow returned to Russia in 1931, he perceived a great change in the atmosphere at the university. Compared with the intellectual freedom he had experienced in Western Europe, the effect of the brutal social reorganizations of Joseph Stalin resulted in an atmosphere surrounding the university that seemed stifling to Gamow. Not only politics but even physics had become a matter of class struggle. It was official party policy that there should be a division between proletarian science and capitalistic science, and the ultimate understanding of nature would result from Marxist philosophy, not from scientists deducing the laws of nature from experiment and observation.

The physics of Einstein as espoused by Gamow and his fellow relativists, including the great Russian physicist Lev Davidovich Landau, was found to be invalid on the basis that it was at odds with the writings of Marx and Engels. Just as three hundred years before when the father theologians had found Copernican cosmology to be at odds with their interpretation of the revealed truth of Scripture, the party found the basic ideas of Einstein's theory of relativity to be contrary to the tenets of Marxism. Just as Galileo was enjoined not to teach Copernicanism, Soviet scientists were forbidden to uphold Einstein's theory of relativity, and those who did were disciplined. Gamow narrowly escaped inclusion in a "condemnation session" ordered by Moscow to straighten out a group of recalcitrant scientists. Impaneled to judge Landau (and Einstein *in absentia*) was a proletarian jury of laborers selected by the party to play the role of the father inquisitors. The machinists, welders, and charwomen on the jury somehow came to the conclusion that space was not curved, and that contrary to Einstein's theory of relativity, an aether did indeed exist and it carried light and electromagnetic waves. Thus, once again, one of the greatest intellectual achievements of the human mind was rejected because it seemed to disagree with something held to be revealed truth. After the session, Lev Landau, the author of the great series of textbooks from which entire generations of physicists all over the world still learn their trade, was dismissed from his teaching position at the Polytechnic Institute.

Gamow was also surprised to learn that of the two equivalent calculational schemes of quantum mechanics, Soviet scientists were ordered to use exclusively the method of Schrödinger, because Heisenberg's method was found to be contrary to the principles of dialectical materialism. This was as ridiculous as saying that it was correct to state that $2 + 3 = 5$ but against the law to write $3 + 2 = 5$.

In search for personal and intellectual freedom, Gamow decided to defect from the Soviet Union, but he was routinely denied exit visas to attend conferences abroad. After a near-fatal, unsuccessful attempt to cross the Black Sea to Turkey in a kayak while on vacation, he finally managed to leave the Soviet Union with his wife in 1933 on the pretext of attending a conference in Brussels. After a couple of temporary visiting fellowships, in 1934 he accepted a position as professor of physics at George Washington University in Washington, D.C., where he stayed until he moved to the University of Colorado in 1956. George Gamow died in Boulder in 1968.

Perhaps because of his exposure to Friedmann from his student days, Gamow never seemed to doubt the big-bang model, and became its leading proponent. But perhaps he wasn't the best salesman for the serious matter of a cosmological theory. Irrepressible, irreverent, irascible, and independent, Gamow stood out among the somewhat gray, somber physicists of the time. If one drove to a hotel where a physics conference was being held and saw in the parking lot an enormous pink Cadillac convertible, with its top down and its fins sticking up over the Ramblers and Studebakers of the other physicists, one knew that Gamow could be found nearby, most likely holding court in the hotel bar. Gamow roared through the otherwise quiet, scholarly world of physics as if still astride his BSA motorcycle with a group of Hell's Angels terrorizing a Sunday School picnic.[5]

Although nearly fifty years later his papers on the big bang seem prescient, at the time they were regarded as flaky. The degree to which his self-promoted image as a Cossack cosmologist contributed to this is unclear. The perception may have been unfair, but the indisputable fact

[5] Gamow even went so far as to write humorous popular books on astronomy and physics, something no serious scientist would think of doing.

is that he did not inspire any astronomer to devote years of research to testing his theories.

About the only things that Abbé Georges Henri Lemaître, the other major adherent of the big-bang model, shared with Gamow were the name "George" and a belief in the big bang. The Belgian-born physicist attended the Catholic University in Louvain after a Jesuit preparatory school. After completing an undergraduate degree in civil engineering, followed by a four-year interruption because of World War I, he received a doctoral degree in mathematics in 1920. Lemaître's career then took a peculiar twist. While Friedmann was developing his version of the big bang in Leningrad, Lemaître entered a seminary in Brussels and was ordained a priest in 1923. He then resumed his career in physics with postdoctoral fellowships at the University of Cambridge and the Massachusetts Institute of Technology before returning to Louvain University in 1927, where he remained until his death in 1966.

Shortly after returning to Louvain, Lemaître discovered the big-bang model, completely unaware that Friedmann had already done so. Despite discouragement from Einstein, Lemaître pursued the idea of the emergence of the universe from what he referred to as the "primordial atom." While Friedmann's and de Sitter's idea of an expanding universe was basically correct, they were worked out essentially as mathematical exercises. Lemaître, on the other hand, explored what we regard today as the fundamental elements of physical cosmology, including worrying about the radiation surviving from the decay of the primordial atom. Perhaps because of his grounding in engineering studies, Lemaître always hoped for some experimental verification of his cosmological theory. His patience was rewarded when, on his deathbed, he learned of the discovery of the background of ancient photons.

As with the work of Gamow, looking back on Lemaître's early work one can see almost all the pieces of modern cosmology. Although Lemaître was a serious scholar, perhaps serious to a fault, his work did not receive the recognition it deserved. As Sir William McCrea of the University of Sussex put it, "All of us who knew him must ever since have wished we had paid better attention to his ideas."

Both Gamow and Lemaître died when I was in high school, so my knowledge of them is only through reading their papers and talking to those who remember them. I cannot say for sure why their work was not more seriously received. Perhaps Gamow's flamboyant style and Lemaître's clerical collar put people off, for few could see the light of their genius. For whatever reason, they were prophets appreciated more today than they ever were in their lifetime.

A MATTER OF DEGREES

Several people realized that there should be a remnant of ancient photons from the big bang, but no one knew what sort of energy the photons had, for in principle they could be anywhere in the **electromagnetic spectrum**. But the location of the photons in the spectrum was the key to determine how one should look for them. For instance, if most of the photons were in the infrared region of the spectrum, then they couldn't be seen with optical telescopes, radio telescopes, X-ray telescopes, and so on, but only with an infrared telescope. Although the big bang predicted that the ancient photons were everywhere in the universe, astronomers had to know *where* in the spectrum to look.

At first, Lemaître thought that the ancient photons might be associated with energetic cosmic rays that rain down on Earth, but this was no more than a guess. It was Gamow who set in motion the events that would lead others to determine the whereabouts of the ancient photons in the electromagnetic spectrum. As so often in the past, the key to unlock an astronomical puzzle was found in a new development in physics.

In the 1930s physicists were just beginning to learn about the structure of the nucleus, and how the nuclei at the centers of all atoms are made out of neutral neutrons and charged protons. It was clear by that time that with sufficient energy two lighter nuclei could be fused together to make a heavier nucleus, or a heavy nucleus could be broken apart into lighter nuclei.

By the mid-1930s, physicists suspected that such nuclear reactions were the energy source of stars, and by the end of the 1930s Hans Bethe and others had worked out the details of the reactions powering the Sun. In addition to learning why the Sun shines, with knowledge of the mechanism for the metamorphosis of one nucleus into another at last it was possible to address the question of why the chemical elements have the abundances they do.

Why is uranium rare and carbon plentiful? Would gold still have its allure if it were as abundant as lead? Questions like these perplexed the greatest scientists of antiquity and the modern scientific era. Isaac Newton invested at least as much time and intellectual energy on alchemy as he did on astronomy. Perhaps he was asking the right questions, but the enterprise was doomed without an understanding of the physical principles involved in changing one element into another. Only with the development of our understanding of the physics of the nucleus came the framework for answering the question of why elements have the abundances they do.

Perhaps the idea of cosmic alchemy appealed to Gamow's free spirit, because he was one of the earliest people to worry about how nuclei were synthesized out of neutrons and protons, a process known as **nucleosynthesis.** His first ideas about the synthesis of the elements in stars were published in 1935, just a year after he arrived in the United States. By 1946 he realized that it was impossible to account for the abundance of all the elements solely by nuclear reactions occurring in the interiors of stars, and he looked around for another locale with temperatures high enough to sustain nuclear reactions.

Before Gamow started on the astro-alchemy path, as early as 1931 Tolman developed a formalism to study the role of radiation in the early universe. Perhaps remembering the lectures of Friedmann more than twenty years previously, or maybe motivated by papers of Tolman or Lemaître, Gamow realized that the early universe was a possible site for the synthesis of at least some of the elements.

If nuclear reactions occurred in the early universe, then by observing the final products of the reactions (the abundances of the light elements today) it is possible to deduce the amount of radiation in the universe relative to the amount of matter, and, by measuring the amount of matter in the universe today, determine the number of ancient photons as well as their location in the electromagnetic spectrum.

The year 1946, when Gamow first proposed the early universe as a site for nuclear reactions, was a watershed year in the development of the big-bang model. In that year Gamow became a consultant to the Applied Physics Laboratory of the Johns Hopkins University in Baltimore, and Ralph Alpher, a part-time student at George Washington University, also joined the laboratory.

Gamow proposed to Alpher that he study the problem of primordial nucleosynthesis for a Ph.D. thesis topic. But at that time Alpher was not an expert in general relativity or the big bang, and where he worked there weren't many cosmologists around with whom to chat. He needed something known as a collaborator. One of the real joys of science is working with someone who shares your excitement about a

Robert Herman (left) and Ralph Alpher (right) in 1949 with the spirit of George Gamow arising from a bottle of ylem, *which is how they referred to the primordial soup.*

subject. Like marriages, some collaborations last a lifetime and others are short-lived. Some collaborations survive passionate arguments at the board while the chalk dust flies, but others don't. Physicists find collaborators in many ways: over coffee, at conferences, over the phone, through "arranged" collaborations, but almost always by happenstance. About the only way collaborators *aren't* found is by taking out an advertisement in a newspaper personal column. If that were a way to find collaborators, in 1946 Alpher might have written:

> Young, white, male physics Ph.D. student seeks a collaborator to share the joy of primordial nucleosynthesis. Must be sensitive to the nuances of general relativity and unafraid to dabble in the big-bang theory. If you would like to build the elements with me, call me at (301) 953-5000. No steady-state cosmologists, please![6]

[6] The steady-state cosmology was the rival model to the big bang at the time. In the steady state the universe was infinitely old and always looked the same as it does now. In the model there was a continuous creation of matter to fill the holes left by the expansion of the universe.

But Alpher didn't have to advertise for a collaborator, because in the neighboring office at the laboratory was Robert Herman, who had been a student of Robertson's at Princeton University and had learned about relativity and the big-bang model there.

So the first research group devoted to the study of the big bang grew in the unlikely environment of a laboratory dedicated to applied physics. Of course, both Alpher and Herman had other responsibilities, so most of their pioneering work was done at night and on weekends. Alpher and Gamow would discuss progress at night in a cafe, Little Vienna, where Alpher would sip coffee and Gamow would drink martinis.

In 1948, just two years after the conjunction of Gamow, Alpher, and Herman, three remarkable papers on big-bang nucleosynthesis by Alpher and Herman appeared, along with Alpher's thesis on the same subject. Combining what they knew about nuclear reactions with big-bang predictions for the rate at which the explosion developed, they were able to calculate the abundances of different chemical elements in the debris of the explosion. One important parameter in their calculations was the density of photons in the universe during nucleosynthesis. If photons were very abundant relative to neutrons and protons, they would impede nucleosynthesis by blasting apart nuclei as soon as they are made.

In the course of the calculations, they realized that the nuclear reactions in the hot primeval fireball would give the observed abundances of the light elements (hydrogen, deuterium, and helium) only for a rather narrow range of the photon density during nucleosynthesis. They could then use this information to predict the *present* temperature of the ancient photons. So within their papers were estimates of the present temperature of the ancient photons.

The number and energy of ancient photons are usually expressed in terms of their temperature. The higher the temperature, the more ancient photons would be around. The temperature of the photons also specifies where they reside in the electromagnetic spectrum. We now know that the temperature of the background photons is 2.726° Celsius above absolute zero (usually rounded off to 3°), or about –455° Fahrenheit. This implies that there are about sixty-five hundred photons in every cubic inch of the universe, and that most of them reside in the radio and microwave region of the electromagnetic spectrum.

The early predictions for the temperature were not very accurate, in part because a lot of needed information about the nuclear reactions that occurred in the early universe was still classified after World War II. Nevertheless, using the information they could find, in 1948 Alpher and Herman estimated that the temperature of the ancient photons would be 5° above absolute zero, reasonably close to the actual value of 3°. Some estimates in their other papers weren't so accurate, running as low as 1° and as high as 28° above absolute zero. But in all their papers the basis on which they estimated the temperature was right on track.

In 1950 Gamow wrote a semipopular account of the big bang for *Physics Today* magazine and quoted a value of 3° above zero for the photon temperature. This prediction is remarkable not only for its uncanny accuracy but also for the fact that Gamow didn't give a clue how he arrived at the value 3° based on predictions that ranged between 1° and 28°.

By the mid-50s there were four published papers by Alpher and Herman discussing the existence of the background of ancient photons: three papers by Gamow, including his *Physics Today* paper, and even a 1949 newspaper article in *The New York Times* that resulted from a press release after Alpher and Herman delivered a paper on the background of ancient photons at a meeting of the American Physical Society.

Remarkably, although the technology was available in the mid-1950s, no one set about looking for the ancient photons until the Princeton effort in 1964 (which was not motivated by Alpher, Gamow, and Herman's work).

Perhaps even more ironic is that the presence of the ancient photons had even been detected over a decade earlier without anyone realizing it. A cyanogen molecule sitting around between the stars occasionally would be hit by an ancient photon, and would live for a while in an excited state. In the late 1930s and early 1940s, three different groups had noticed that molecules of cyanogen detected in the interstellar medium behaved as if they were immersed in a bath of photons with a temperature between 2° and 3°. This fact was noted, but its significance not appreciated.

THE SOUNDS OF SILENCE

The instrument that would be used in the "discovery" of the ancient photons was the twenty-foot antenna built by Art Crawford of Bell Labs to detect signals from the Echo balloon. The Echo idea was that a signal sent from California would bounce off a balloon orbiting Earth and be detected in New Jersey. This doesn't sound too exciting in this age of backyard satellite dishes, but it was state-of-the-art in the late 1950s and early 1960s. After Project Echo, the Bell Labs antenna was used to receive a beacon from Telstar, the first true communications satellite.

By 1964 the antenna was no longer needed for satellite communications, and it was turned over to Penzias and Wilson for radio astronomy work. While predictions of ancient photons sat in dusty journals, in New Jersey these two watchers of the sky would use the antenna to see something others had seen before but ignored. Although at first blind to what they were seeing, Penzias and Wilson could not close their eyes to the ancient photons for long.

Penzias and Wilson were interested in making an *absolute* measurement of the radiation from a supernova remnant known as Cas A as well as other sources, not just a *relative* measurement as others had done. To make an absolute measurement is much more difficult than to make a relative measurement. Say you have two glasses of water, one at room temperature and the other just a couple of degrees warmer. It is pretty hard to tell the temperature of the water in one of the glasses just by putting your finger in it, but it is quite easy to tell if the water in one glass is just slightly warmer relative to the water in the other. Whereas others had performed relative measurements, Penzias and Wilson wanted to measure the exact amount of radiation from Cas A, not just if it was hotter or colder relative to some other source. They decided to

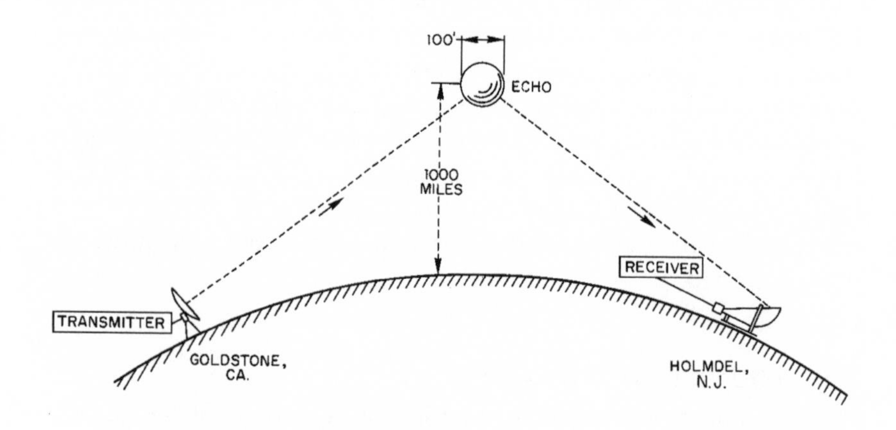

The horn antenna was originally constructed as part of Project Echo. A signal from Goldstone, California, was bounced off of a hundred-foot Mylar balloon in orbit and detected in Holmdel, New Jersey.

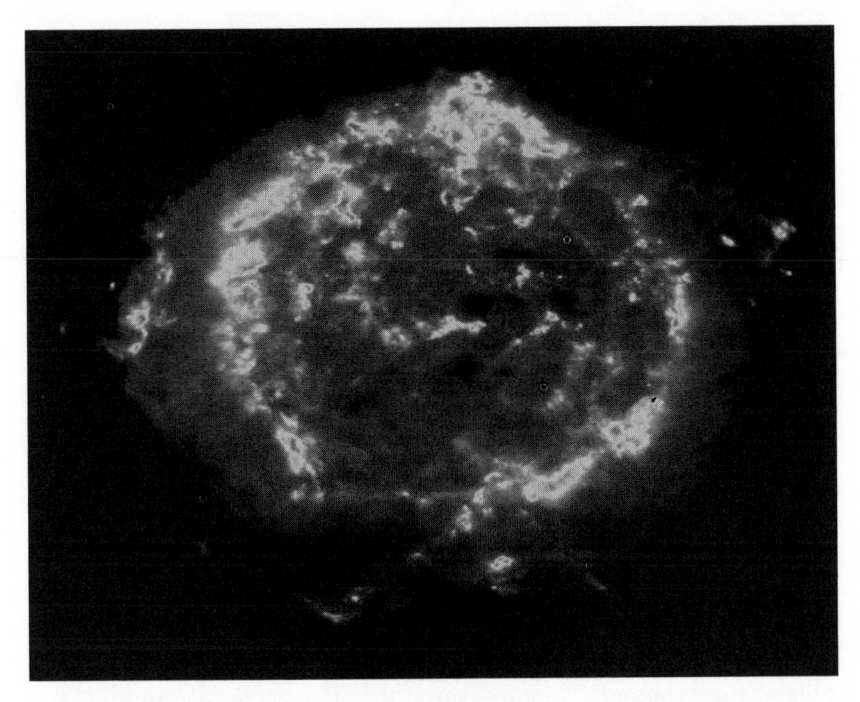

A radio picture of Cas A, a remnant of a supernova explosion seen in the constellation Cassiopeia. It is about ten thousand light-years away and has a radius of about fifteen light-years. Partly because of its proximity it appears to us as the brightest source of radio emission.

determine the absolute flux by comparing the signal from the antenna when it was pointed at Cas A with the signal from a source at a known temperature, in their case liquid helium at about 5° above absolute zero. It was as if they would point their finger first toward Cas A, then toward the liquid helium, and see which one was hotter.

In addition to the radiation from Cas A, tumbling into the horn antenna on Crawford Hill were photons from all sorts of spurious sources: from Earth's atmosphere, from the walls of the antenna, from the ground, and from nearby New York City. Added to the signal of the photons entering the antenna from outside was noise in the plumbing through which the photons passed, just like there is noise in any radio receiver. To make an absolute measurement of the radiation from Cas A, Penzias and Wilson had to separate the signal of the photons from the source from the stray unwanted photons in their detector.

Before Penzias and Wilson used the antenna for astronomy, engineers and physicists at Bell Laboratories had attempted to measure the amount of spurious radiation detected by the horn antenna. They thought they could accurately separate out the amount of radiation from the atmosphere, from the antenna walls, from the ground, and from the plumbing. When the accounting was done, however, they saw about 15 percent more radiation from the antenna than they expected. They made note of this curious fact, but they did not realize that the extra radiation was a discovery worthy of a Nobel Prize. Although the 15 percent difference was curious, it was not such a striking discrepancy that it couldn't be ignored, for after all, *the most difficult thing to discover is something for which you are not searching.*

Because Penzias and Wilson wanted to make absolute measurements, they couldn't ignore the 15 percent effect. To isolate the noise in the amplifiers, switches, and the rest of the plumbing from the true signal from the sky, they constructed what is known as a "cold load," which was the bottle of liquid helium at 5° above absolute zero.

They thought that the only significant sources of spurious radiation entering the detector would be from the atmosphere and from the walls of the antenna.[7] They measured the effective temperature of the atmosphere to be about 2.3° above absolute zero, and they expected the walls of the antenna to contribute about another degree or so. Before measuring the radiation from Cas A they tested the equipment by pointing the antenna to another region of the sky where there was no expected extraterrestrial signal. In addition to the small noise from Earth's atmosphere and the walls of the antenna, Penzias and Wilson expected to hear the cosmic sounds of silence.

On the morning of May 20, 1964, when the antenna first pointed to a region of the sky without an apparent source of radiation, they heard static rather than silence. They had expected the sky to appear at a temperature of between 3° and 4° above absolute zero due to radiation from the antenna walls and from the atmosphere. Since the liquid helium was at 5°, they expected the antenna signal to be smaller than the signal from the helium. But on their very first day they saw that the sky was hotter than the helium. That meant that in addition to the stray radiation from the atmosphere and from the walls of the antenna there

[7] They could isolate the part of the signal due to the atmosphere by pointing the antenna at different angles above the ground and varying the thickness of the atmosphere through which they looked.

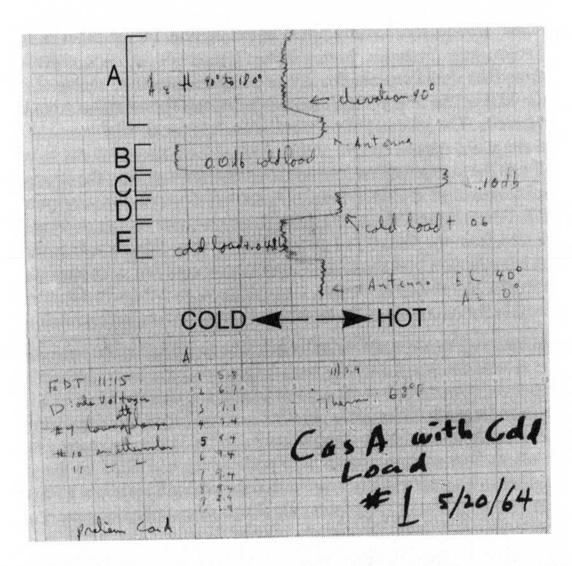

The first unmistakable signal of ancient photons is shown in this picture of the chart recording of May 20, 1964, comparing the signal from the antenna with the signal from the liquid helium. The intensity of the signal increases to the right. At the top in the first part of the graph in the region marked A *is the signal from the antenna, followed in part* B *by the signal from the cold liquid helium. Penzias and Wilson were surprised to find that the liquid helium at 5° above absolute zero was colder than the sky. In parts* C, D, *and* E *they added a little bit of noise to the signal from the liquid helium to simulate a higher temperature cold load. In the part marked* E *they found that the signal from the antenna was equivalent to about 7° above absolute zero. With 2° from Earth's atmosphere, and another 2° from the walls of the antenna, that left 3° unexplained—the 3° background of ancient photons.*

was another source of radiation entering the antenna. We now know Penzias and Wilson had found the ancient photons, but they thought they had only found a problem.

Penzias and Wilson thought this problem was a real show stopper. Their planned research project of measuring the absolute flux of astrophysical sources seemed at risk if there was an additional source of noise in their equipment that they couldn't account for. Over the next months they carefully repeated the experiment after checking out every component of the antenna, even going so far as to remove two pigeons who had made their home in the antenna. Still the signal persisted. The pigeons were gone, but they had left behind on the bottom of the antenna something referred to by Penzias and Wilson as a "white dielectric substance." Perhaps Penzias and Wilson knew that either they had made one of the greatest discoveries in the history of cosmology or they were making the most accurate measurements of the conductivity of pigeon droppings that anyone could ever want. So they climbed into the antenna and scraped away the guano.[8]

Penzias and Wilson were scratching their heads about the source of the radiation when suddenly the direction of their work, the ongoing project at Princeton, and the work at Johns Hopkins accidentally converged. The unfolding of the events has been told often, but the irony is just too striking to resist retelling.

In January 1965, more than six months after seeing the signal of ancient photons, Penzias happened to mention the unknown source of radiation while speaking on the phone to Bernard Burke, another radio astronomer. At the time Burke was working at the Department of Terrestrial Magnetism of the Carnegie Institution in Washington, but he had heard from a third colleague, Ken Turner, about a talk given by Jim Peebles of the Princeton collaboration who were in the process of mounting their own search for ancient photons while Penzias and Wilson were wondering what they were seeing. In the seminar Peebles discussed the possibility of a background of ancient photons. Just to complete the circle, Peebles had presented the seminar at the Johns Hopkins University, of all places, where a decade previously (unknown to Peebles) Alpher and Herman, inspired by Gamow, had developed the theory of primordial nucleosynthesis, which led to their own prediction of the background of ancient photons. (Neither Alpher nor Herman was around to hear Peebles—both had left to work in industry.)

Finally, after the Peebles to Turner to Burke to Penzias connection, the different dancers realized that they were all stepping to the same tune. In May 1965, Penzias and Wilson published their results in the *Astrophysical Journal*, alongside a paper by the Princeton group interpreting the results. Later that year Dicke and Roll completed their measurement of the photon background and confirmed the discovery.

[8] This is yet another illustration of the fact that the process of discovery is not a sterile exercise one does while sitting under an apple tree. Usually some sort of dirty work is involved in discoveries. Of course, the dirty work of a theorist is usually cleaner than the dirty work of an experimentalist.

Immediately several people realized that the excitation of cyanogen was due to the ancient photons.

If someone had scripted the story of the discovery of the ancient photons in this way, most likely it would have been deemed too implausible even for television.

What are the lessons in this oft-told tale? Perhaps the most important moral of the story is that understanding the universe is no easy task. Nobel laureates Penzias and Wilson saw the photons but at first didn't know what they were. In 1950 another Nobel laureate, Gerhard Herzburg, had written about the excitation of cyanogen molecules that it implied "a rotational temperature of 2.3 degrees [above absolute zero] which has of course only a restricted meaning." It wasn't the temperature that was restricted, but their understanding of the meaning of the observation. The other people involved were world-class physicists: Robert Dicke was one of the great experimentalists of the time, Jim Peebles is today one of the foremost cosmologists, and Wilkinson continues to make important measurements of the properties of the background photons. After the discovery people realized immediately that they should have taken Gamow and Lemaître more seriously. Even now the prescient work of Alpher and Herman has not been properly recognized.

All of these great scientists were circling around the discovery, occasionally tripping over it, while the universe patiently waited for our wits to grow sharper. We are all, in our own way, blind watchers of the sky.

TEN

Primordial Soup

When visiting a science museum or planetarium, I always stop to look at displays of antique astronomical instruments, sometimes spending hours trying to imagine what it was like to be an ancient watcher of the sky. After studying the instruments for a while, I like to wander around and try to locate representations of ancient views of the universe. The views are sometimes displayed in the form of enlargements of figures from old books, but if I am really lucky the museum will have some of the physical models once built to depict the apparent motions of stars and planets. Occasionally a passerby will express amusement at the strange visions of the universe depicted in the models, or at the primitive nature of the instruments, but my impressions are quite different. I can easily spend an entire afternoon in quiet admiration of the tools and models of astronomers of the past, for I truly believe that in the age-old quest to understand the universe, the dedication, cleverness, and imagination of those ancient watchers of the sky were no less than ours. It is all too easy to scoff at their Earth-centered cosmology with eccentrics, epicycles, and the like, but I believe the culmination of their labors was a tremendous intellectual accomplishment, in many ways as beautiful and remarkable as any of the seven wonders of the ancient world.

In one sense the task of a modern cosmologist is easier than the job of our ancient brethren because of the array of modern scientific tools at our disposal. Just as the tools of ancient astronomers were a realization of their desire to understand the universe, ours also represent a similar yearning. I imagine that some biologists treasure microscopes, and some chemists might even love the sight and feel of test tubes, but I think that astronomical instruments are really something special. Some might regard a telescope as a cold metal tube containing a few lifeless pieces of glass, but reflected in the mirror along with the light of stars and galaxies is a noble human aspiration. The telescope is a way by which we concentrate the light from the distant reaches of the universe in order to bring the understanding of our surroundings into sharper focus. For this reason I feel a strange sense of reverence when looking at a telescope. Although I feel that there is something noble about space telescopes and other marvelous machines used by modern astronomers, I am no less moved by the inventions of our predecessors.

All scientific tools, whether ancient or modern, are conceived and constructed within a framework of a picture of the universe. A complicated modern instrument is not simply a jumble of metal, wires, and silicon chips, but a physical realization of a mental picture of the universe. In their own way, ancient instruments are also a realization of a view of the universe. All experimental apparatus reflects a schema in the mind of its makers, and in turn a successful instrument will lead to a refinement of the model on which it was based.

The human mind has always been the greatest tool of discovery, but the imagination has usually been limited by the ability of instruments to reveal the wonders in the sky, for to imagine what cannot be seen is an exceedingly rare ability. Philosophers and scientists debated the constitution of the heavens for millennia, but until Galileo turned his optic tube toward the sky no one had imagined that there were stars invisible to the naked eye or there were moons orbiting other planets.

Before the invention of the telescope, astronomers could see the sky only with the unaided eye. But the marvelous physiology of the human eye evolved for the purpose of seeking food and avoiding predators, not for looking at dim points of light in the dark night sky. In spite of this handicap ancient astronomers fashioned a number of ingenious observational tools to augment the comparatively limited optics at their disposal.

Of all the astronomical apparatus of antiquity, my favorite is the astrolabe. The planispheric astrolabe is a disk of metal or wood with

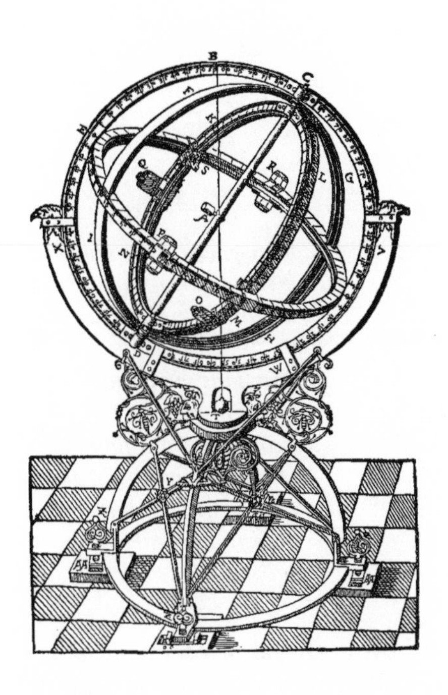

*An armillary sphere, representing the motions of the planets and stars about a stationary Earth (*armilla *is Latin for "ring"), is a three-dimensional model of the universe. This armillary sphere was one of many built by Tycho Brahe. In his book* Astronomiæ Instauratæ Mechanica *of 1598, he describes how this one could be disassembled easily and moved in case a patron decides to stop support for astronomy research and it is necessary to move to another country. As Tycho said, "... everywhere the Earth is below, and the sky above, and to the energetic man, any region is his fatherland."*

which astronomers could locate stars, measure the altitude of stars or the Sun above the horizon, tell the time of day, determine directions before the invention of the compass, or find the latitude on Earth of the observer. The astrolabe is in fact an analog computer, and may well have been the world's first scientific instrument.

The development of the astrolabe is traditionally credited to Greek astronomers, either Hipparchus (ca. 150 B.C.) or Apollonius of Perga (ca. 250 B.C.), but the true identity of the individual (or, more likely, individuals) who invented the astrolabe is likely forever lost to time. Ptolemy probably used one to sight the stars when he perfected his cosmological model in the second century of the Christian Era. There are no surviving Greek astrolabes; the earliest instruments we have today are Islamic astrolabes from the second half of the tenth century. Islamic astronomers refined the astrolabe, in part because in addition to its use in charting the heavens it was a wonderful tool that could be used to determine which time of the day to pray, and how to find the direction of Mecca to face in prayer. Many of the Islamic astrolabes are exquisitely decorated, precisely engineered works of consummate craftsmanship that would find an honored place in an art museum as well as a science

A planispheric astrolabe is in some sense a two-dimensional stereographic projection of the three-dimensional armillary sphere. This remarkable and versatile intrument could be used to tell time, find the position of heavenly bodies, determine latitude, and be used in terrestrial surveying.

museum. But every tool, no matter how beautiful or useful, eventually becomes obsolete; and although it continued to be fashioned and used in the East even as late as the nineteenth century, by the end of the seventeenth century the astrolabe was supplanted in the West by other instruments.

Although some might dismiss it as a primitive tool, the astrolabe is more than just mere pieces of metal, for engraved on its delicate parts is in some respect an elaborate picture of the heavens. Depicted by the intricate movements of rings of brass is a complete representation of a model for the motions of the heavenly bodies, which in turn was possible because people understood the motions in the sky based on a cosmological model. Construction of the instrument would have been impossible without a model of the sky (along with some sophisticated mathematics), and, in turn, its use led to further information about the sky. In this respect it is the same as our modern instruments.

I don't know why I am so attracted to astrolabes. Perhaps I like them because their circular shape reminds me of a tool used by modern cosmologists, a tool that I employ in my research—the particle accelerator.

Particle accelerators, sometimes referred to by the romantic but misrepresentative name "atom smasher," are used to accelerate subatomic particles to velocities near the speed of light, and to smash them together, creating a fireball of energy contained within an incredibly small collision region.

Just as few people have seen an astrolabe, or could identify what it is if they did see one, not many people have seen a particle accelerator or would know what it is even if looking down on one. But looking out of airliner windows flying in and out of Chicago, I often observe the circular shape of the accelerator of Fermi National Accelerator Laboratory, located just eighteen miles southwest of O'Hare Field. I can easily identify the outline of the accelerator; its circular shape seems to jump out of

The particle accelerator at Fermilab, located in Batavia, Illinois, about thirty miles west of downtown Chicago. The large ring outlines an underground tunnel that contains powerful magnets shown in the lower picture. The magnets are used to steer the particles around the ring and to guide them to the collision regions. While ancient astrolabes could be held by hand, the radius of this astronomical tool is one kilometer (0.62 miles, for those nonconversant in metric-speak).

the surrounding cornfields. But I have a big advantage, because I know what to look for, which more often than not is the key to successful observations. I am sure that the ring escapes the attention of most passengers, who are probably more concerned about their impending landing at the world's busiest airport.

Just as the astrolabe was a tool used by ancient astronomers, the particle accelerator is a tool used in modern times. The astrolabe has a representation of the universe engraved on it, while the particle accelerator is a view of the universe engraved into Earth. The astrolabe allowed ancient astronomers to reach out in space into the heavens and chart the motions of the planets, while the particle accelerator enables modern cosmologists to reach back in time into the early universe and understand its evolution.

Particle accelerators are built to probe the inner structure of matter. They reveal the inner workings of the universe on the smallest distance scales we can examine. In this respect they are giant microscopes. Perhaps many people find it surprising that something that functions as a microscope can also be used as a telescope to explore the universe on the largest distance scales, or as a time machine to study the early universe. But the accelerator is an important instrument for anyone interested in understanding the earliest moments of the universe, because in the fireball of energy produced by the collision of high-energy particles, it is possible to taste the primordial soup of the big bang.

AS IT WAS IN THE BEGINNING

In the long history of civilization, every culture has developed its own individual view of the universe. In 1945 the great American anthropologist George P. Murdock listed sixty-eight common characteristics of every culture known to history and ethnography. Along with marriage, puberty customs, hair styles, and bodily adornment, every culture has had a shared belief in a cosmological model. Whether passed down as legends spoken by elders around campfires, integrated into religious ceremonies, taught in schools, or transmitted through books, lectures, or even television, a common view of the universe is promulgated throughout society.

Anthropologists note that cosmological models have been as diverse as the cultures that conceived them. But in spite of the differences between them, there are some commonalities. For instance, most cosmological traditions ascribe an anthropomorphic origin to the universe and speak of its "birth" or creation. This view of the universe is the one familiar to most of us who are a product of modern Western civilization, because our worldview is dominated by the Judeo-Christian biblical legend found in Genesis. But not all cultures believed in a beginning for the universe, and many religious traditions are without a legend corresponding to Genesis.

An example of a no-creation cosmology is the Jainist tradition, which arose in India sometime in the sixth century B.C. and holds that there was no beginning to the universe. Expressing this philosophy, the Jain teacher Jinasena (ca. A.D. 900) wrote:

> Some foolish men declare that Creator made the world.
>
> The doctrine that the world was created is ill advised, and should be rejected....
>
> Know that the world is uncreated, as time itself is, and is without beginning and end.

Until relatively recently in the history of civilization, creation stories were based on myth, and they could either be accepted or rejected, but never tested. Following the work of Copernicus, Tycho, Kepler, Galileo, Newton, and others responsible for the emergence of modern science, however, the Western view of the universe has been grounded in science rather than religion or philosophy. Today it is scientists who are the keepers of our cosmological worldview, and scientific models are subject to falsification.

Although the development of astronomy and physics in the seventeenth, eighteenth, and nineteenth centuries did so much to deepen our understanding of the solar system and our galaxy, it could not address the question of the origin of the universe because there were no observations or theoretical framework from which one could infer the age of the universe or even address the question of whether the universe is timeless.

That situation changed after Hubble discovered the expansion of the universe. Although scientists (including Hubble himself) were slow to accept the profound implications of his discovery of the expansion, the discovery of ancient photons by Penzias and Wilson clinched the case for the big-bang origin of the universe. Finally there was a scientific basis on which to address the question of origins.

The big-bang picture is not that of a static universe, but one with galaxies moving apart in expansion. If the universe is expanding today with galaxies receding from each other, it must have been denser in the past. If our moving picture is run backward in time, it suggests that the universe emerged from a state of extremely high density some time in the past. Because of the uncertainty in the measurement of the present rate of expansion of the universe due to the unreliability of our determination of the distances to very distant objects (again, determining the distance to the things we see in the sky has always been a fundamental problem in astronomy), the expansion age of the universe is uncertain. Entire sessions during cosmology conferences are devoted to passionate arguments about whether the age of the universe is ten, fifteen, or twenty billion years. Whatever the true value (my money is on fourteen billion years), the big-bang model predicts that the universe we see today emerged from a high-density state a finite time in the past.

However long ago the expansion started, the early universe was a much different from the one we see now, for not only was it denser in the past, but it was hotter.

Today the temperature of the background photons is known very accurately: between 2.716° and 2.736° Kelvin above absolute zero. This temperature is usually rounded off to 3° Kelvin (or about –270° Celsius, –455° Fahrenheit).[1] But the big-bang model also predicts that as the moving universe picture is run backward toward the beginning of the reel, the temperature of the early universe was larger. In other words, the expanding universe started hot, and cooled as it expanded. The mathematical machinery of the big-bang model can predict the temperature of the universe at any time during the expansion.

The largest uncertainty in the calculation of the temperature during the history of the universe is the present age of the universe. Some of my colleagues might argue for a younger universe, and some might defend a larger value for the age, but let's ignore the cacophony of cosmologists and adopt fourteen billion years for the present age of the universe.

Using the big-bang model, we can journey back in time and imagine what the universe was once like. When Earth formed some 4.6 billion years ago, the universe was already more than 9 billion years old. But the temperature of the universe has not increased much in the last

[1] Although there are regions of the universe warmer than 3° above absolute zero, noticeably in the vicinity of stars, most of the volume of the universe is empty space containing the background photons. For this reason it makes sense to speak of 3° as the temperature of the universe.

4.6 billion years; it was only about a degree warmer than today, or about –269° Celsius, when Earth formed. The universe at that time closely resembled the universe we observe today.

Because of the finite velocity of light, when we look at distant objects we are actually observing them as they were when the universe was younger. When we look at the Sun, we observe it as it existed eight minutes previously, because it takes light from the Sun that long to reach us. If we look at the bright star Sirius, we see it as it existed eight and one-half years ago, because Sirius is 8.5 light-years distant. When Hubble measured the distance to the nearby galaxy Andromeda, the light from the stars in that galaxy traveled over two million years before it burned into his photographic plates, so his image of Andromeda showed the galaxy as it existed over two million years ago.

Quasars are the most distant (hence the oldest) astrophysical objects we can observe. The light we now observe from the most distant quasar was emitted some thirteen billion years ago, when the universe was only about one billion years old and had a temperature of –257° Celsius. Quasars probably signify the earliest epoch of galaxy formation, and may well be the most distant individual objects we will ever see. Although the 13° Celsius (23° Fahrenheit) change in the temperature of the universe since quasars formed thirteen billion years ago is often exceeded by the daily temperature change on Earth, the universe was a much different place then from now. For instance, the way matter was distributed was very different.

We would not find the average mass density of a typical star like our Sun very unusual; it is between the density of water and the density of rock. But we would find the average mass density of our galaxy very unfamiliar. There is so much space between stars that if the mass of all the stars in our galaxy were spread evenly throughout the galaxy rather than clumped into stars, the mass density of the galaxy would be about one million million million million times less than the density of water. The average mass density in our galaxy is much smaller than that found in the best vacuum we can produce on Earth. There is also a lot of empty space between galaxies. If the mass of all the galaxies were spread evenly throughout the universe, the average mass density would be a million times smaller still. In other words, today galaxies are a million times larger than the average density of the universe, and stars and people one million million million million times larger still.

But well before the quasar epoch, galaxies and stars did not exist, and matter was spread smoothly around the universe. Before the time quasars formed, regions that would eventually form galaxies, stars, and people had a density only a *few percent* larger than the average density of the universe, rather than one million...million times larger. Over billions of years the gravitational attraction of these regions of slight overdensity accreted surrounding matter, eventually growing to become the structure we observe. Although any tiny irregularities in the density of matter will inexorably grow, the nature of the resulting structure depends on the amplitudes and sizes of the irregularities. But to learn the nature of the irregularities, we must study the early universe much, much earlier than the quasar epoch.

Going back farther in time, we expect that the temperature of the universe was a pleasant 22° Celsius (72° Fahrenheit) when the universe was about 12.5 million years old. Of course, there is no cosmological significance to the epoch in the history of the universe when the average temperature was comfortable for human habitation. Earlier than about 3.3 million years after the bang, the universe was too warm for comfort. In fact, it was hotter than hell (assuming that the temperature of hell is about the boiling point of brimstone, 445° Celsius). This era might have theological implications, but nothing of interest to cosmologists occurred at this time; the universe seemed to pass uneventfully through the temperature of hell.

The temperature of 3,300° Celsius does mark an important cosmological event. When the temperature was higher than this, the universe was so hot that electrons would have been stripped from atoms and all of the atoms in the universe would have been ionized. Today there are atoms in the universe, but before the ionization era any atoms would have "melted" and the universe would have consisted of a plasma of nuclei and electrons. After the universe cooled enough for the electrons to combine with nuclei and form atoms, radiation was free to roam. Earlier than 300,000 years after the bang, however, the ionized universe was opaque to radiation. It was at the time atoms first formed that the ancient photons last interacted with matter. Just as when we observe a distant object we are seeing it as it existed when the light was emitted, when we detect the ancient photons we are seeing the universe as it existed nearly fourteen billion years ago when it was only 300,000 years old.

Because radiation cannot travel through the ionized universe, we cannot look out in space and look back to a time before the ionization era, no matter how large a telescope we build. We just cannot see through an ionized plasma, and we must find another way to observe the early universe.

We study the very early universe by making a little piece of it in the laboratory. Although the universe at the ionization era was hot and dense compared with the present universe, the conditions are not extreme compared what what we can produce on Earth. In fact, the density of matter at ionization was quite small, one thousand million million million times less dense than water. And while a temperature of 3,300° Celsius is uncomfortably hot, it can easily be produced in terrestrial laboratories. From laboratory experiments we have a complete understanding of the laws of nature that were important under the conditions of ionization, and it is with confidence that we can speak of what the universe was like at that time.

In our typically egocentric view of the universe, we usually imagine that the stuff of which we are made (neutrons, protons, and electrons) are the only important things in the universe. But there is also the invisible background of ancient photons. While photons do not have mass, they do have energy, and using the famous $E = mc^2$ equation we can speak of an equivalent mass density of the background radiation. We know that as the universe expands, the radiation cools and the equivalent mass density decreases. If the moving picture of the expanding universe is reversed, then the radiation energy density was larger in the past. Although today the energy density of radiation in the universe is much smaller than the mass density of the matter, earlier than thirty thousand years after the bang the density of radiation was larger. In the very early universe, the matter we think of as so important played a negligible role in the dynamics of expansion, and the universe was dominated by radiation.

Just as the atoms melt into electrons and nuclei if the temperature of the universe is larger than 3,300° Celsius, the atomic nuclei cannot stand the heat if the universe is hotter than 10,000,000,000° Celsius. Before the universe reached this temperature, about a second after the bang, any nuclei present would have melted into their constituent particles, neutrons and protons.

Only after the universe cooled below 10,000,000,000° Celsius could elements produced by nuclear reactions survive without being immediately blasted apart by radiation. But only three minutes later the temperature had dropped to one billion degrees, and it became too cold for nuclear reactions to occur. So the universe had only three minutes to cook the elements from the primordial neutrons and protons into nuclei before it became cold enough that nuclear reactions shut down.

In this era the matter density was still not very high, about the density of rock. The temperature of the universe when the nuclear reactions occurred was enormous by normal experience, but energies corresponding to those temperatures have long been produced in nuclear physics experiments. Since we can reproduce those temperatures and study how neutrons and protons combine and make heavier nuclei in a process known as **nucleosynthesis**, we can expect to make reasonable predictions about what occurred in the universe in this era. Although it occurred fourteen billion years ago, we can speak about it as if it happened just yesterday, because in a sense it did, for just yesterday the temperature of the universe during the nucleosynthesis era was reproduced and studied in terrestrial laboratories.

HISTORY OF THE UNIVERSE (IN BRIEF)

EPOCH	AGE	TEMPERATURE
REPRODUCIBLE IN TERRESTRIAL EXPERIMENTS		
TODAY	*14,000,000,000 yrs.*	*–270 C*
EARTH FORMS	*9,400,000,000 yrs.*	*–269 C*
OLDEST QUASARS FORM	*1,000,000,000 yrs.*	*–257 C*
ROOM TEMPERATURE	*12,500,000 yrs.*	*22 C*
HOT AS HELL*	*3,300,000 yrs.*	*445 C*
ATOMS MELT	*300,000 yrs.*	*3,300 C*
RADIATION ERA	*30,000 yrs.*	*14,000 C*
NUCLEOSYNTHESIS ENDS	*3 min.*	*1,000,000,000 C*
NUCLEI MELT	*1 sec.*	*10,000,000,000 C*
NEUTRONS/PROTONS MELT	*0.0001 sec.*	*1,000,000,000,000 C*
PRIMORDIAL SOUP	*0.000000000004 sec.*	*3,000,000,000,000,000 C*
START OF THEORETICAL SPECULATION		
ELECTROWEAK UNITY	10^{-13} sec.	10^{16}C
STRONG UNITY	10^{-35} sec.	10^{28}C
QUANTUM GRAVITY	10^{-43} sec.	10^{32}C

*In the absence of direct information, hell is assumed to be about as hot as the boiling point of brimstone.

Some interesting developments in the history of the universe. The temperatures are given in Celsius. To convert the temperature to Fahrenheit, multiply the above numbers by 1.8 and add 32.

An important big-bang prediction is the chemical composition of the universe after primordial nucleosynthesis. If we start with an expanding universe of neutrons and protons, about three minutes later the neutrons and protons will be cooked into a soup of nuclei that is about 76 percent hydrogen and 24 percent helium, a few parts in a million of other isotopes of hydrogen and helium, and about one part in ten million lithium. Conspicuously absent from the primordial mix were the heavier elements, like carbon, oxygen, calcium, iron, and all the other elements so crucial for life.[2] It turns out that this predicted mix of elements is just about exactly what astronomers think was around before the first generation of stars formed and cooked the primordial elements into heavier nuclei in their nuclear furnaces. The prediction comes from a true interplay between the big-bang model and nuclear physics. If the universe expanded more slowly than predicted, a smaller fraction of helium would have resulted; a slightly faster expansion would have resulted in too much helium. If the laws of nuclear physics had been different fourteen billion years ago, a different mix of elements would have been produced. We can make a prediction about the result of events that occurred fourteen billion years ago, and since the observations agree with the prediction, we have some confidence that the model is correct.

But neutrons and protons are not the beginning of the story. In the twentieth century we learned that atoms exist, but despite the fact that the word *atom* derives from the Greek word for "indivisible," they are not elementary. Atoms are not indestructible, but can be broken apart into electrons and nuclei. Not only can atoms be split, but we found in the 1930s that nuclei can be divided further into neutrons and protons. In the 1960s we discovered that neutrons and protons are not elementary, but they are made of quarks and gluons. We know the universe was once hot enough to melt atoms, because the ancient photons are a relic of that era. We know the universe was once hot enough to melt nuclei, because we see the elements that were produced in that era. If we continue to reverse the moving picture of the expanding, cooling universe, it should have once been hot enough to melt the neutrons and protons into their constituent particles—quarks and gluons.

Individual neutrons and protons first appeared in the universe about 0.0001 seconds after the bang. At this time the temperature and

[2] These elements were produced billions of years later during *stellar* nucleosynthesis.

density of the universe were so high that neutrons and protons were touching, and for a brief fraction of a second the entire universe was as one giant nucleus. The density of this giant nucleus was so great that the entire mass of the Sun occupied a sphere only a few miles in diameter, and all the mass in the Milky Way was contained in a sphere not much larger than Earth. We do not yet have a complete understanding of the physics of matter at such extreme density, but many of its most important properties can be calculated, and the calculations checked by laboratory experiments.

Before the nuclear-density era, the universe was so hot and dense that individual neutrons and protons could not exist as such, but melted into their constituent particles, quarks and gluons. So we believe that the universe younger than 0.0001 seconds was dense soup consisting mostly of quarks and gluons.

IS NOW

In one Chinese creation myth (ca. third century A.D.), the birth of the universe was the result of the death of the god P'an Ku.

> The world was never finished until P'an Ku died. Only his death could perfect the universe. From his skull was shaped the dome of the sky, and from his flesh was formed the soil of the fields; from his bones came the rocks, from his blood the river and the seas; from his hair came all vegetation. His breath was the wind, the voice made thunder; his right eye became the moon, his left eye the Sun. From his saliva or sweat came rain. And from the vermin that covered his body came forth mankind.

Of course, this creation myth seems fanciful to us now (some people still can't deal with the fact that we are related to chimpanzees, let alone descended from vermin), but the tale must have seemed believable at the time. Our picture of the universe makes sense to us (at least to me), but we might have a hard time convincing a Chinese peasant from the third century that fourteen billion years ago the universe was a hot, formless primordial soup of quarks and gluons, and embedded in this soup were small inhomogeneities that would eventually grow to become galaxies, stars, planets, and people. So what makes the big-bang model more than just another creation myth?

The picture of the big bang is different in kind, not in degree, from other pictures such as the P'an Ku myth. One is a scientific model, whereas the others are just myths.

There are three major attributes of the big-bang model that separate it from previous pictures of the universe. The first is that the big-bang model can make predictions. An example of such a prediction is the primordial abundance of the elements. The ability of the big-bang model to make such a prediction illustrates how fertile the model is.

The big-bang model was not conceived to explain the origin of the elements. It was developed in the 1920s by Friedmann and Lemaître as an application of Einstein's new theory of gravity. In the 1930s Robertson and Tolman realized that the temperature of the early universe would be higher in the past than it is now. In the 1940s and 1950s, Gamow, Alpher, and Herman realized that there should be a background of ancient photons, which was later stumbled on by Penzias and Wilson in the 1960s. Shortly after the discovery of the ancient photons, Robert Wagoner, William A. Fowler, and Fred Hoyle refined the calculations of Alpher and Herman and calculated the relative abundances of the elements produced in the big bang. The predictions were subsequently confirmed by astronomers. Today there is a continuing cycle of refinement of the calculations and more accurate observations.

The second reason to prefer the modern model is that the conditions of the big bang can be recreated and tested by experiments in terrestrial laboratories. No one can reproduce the death and decay of P'an Ku, but the primordial soup can be made and studied today.

Of course, the two reasons are connected in a deep and profound way. It is the prediction of the existence of two relics of the early universe, the microwave background and the primordial abundance of the elements that gives us confidence in the big-bang model, and it is the input of the fundamental laws of physics as discerned by terrestrial experiments that allow us to make the predictions.

Well, so what? We have a model for the beginning of the universe, but what do we do with it? The answer to this question leads to a third and even more important reason that the big-bang model is preferable to other models: it has the potential of leading to an even deeper understanding of the universe than we have today. After all, why do we want a model in the first place? If we recall the great controversy of the

Ptolemaic versus the Copernican model, the ultimate reason the Copernican view triumphed was not that it was a better model for making predictions, but that it led to a deeper understanding of nature and the connection of physics and astronomy. When writing about Galileo's great *Dialog on the Two Chief World Systems*, Einstein said:

> In advocating and fighting for the Copernican theory, Galileo was not only motivated by a striving to simplify the representations of the celestial motions. His aim was to substitute for a petrified and barren system of ideas the unbiased and strenuous quest for a deeper and more consistent comprehension of the physical and astronomical facts.

There are many astronomical facts that cry out for a deeper and more consistent explanation. I could rattle off a dozen, but let me just mention two. We know that there is matter in the universe, but we don't know why. A universe with just radiation, without any matter at all, is perfectly consistent with the known laws of physics. So why is there matter? A second mystery has to do with the way the matter is arranged. We know that the distribution of matter in the universe before the quasar epoch was very smooth, but not perfectly smooth. Why wasn't it either perfectly smooth or very lumpy? Both the perfectly smooth model and the very lumpy model are consistent with the laws of physics as we know them. Why were there any imperfections in the fabric of the universe, and why were they small?

As a cosmologist I am lucky, because not only are there fundamental issues, but there is a model that holds promise for providing a framework within which to address these problems.[3]

Most cosmologists believe that the solution to at least some of the unanswered questions of astronomy will be found if we understand the history of the early universe. Perhaps the existence of matter and the lumpiness of the distribution of matter are relics of events that occurred in the universe earlier than 0.0001 seconds after the bang. So the push to

[3] I am really lucky on the first count because if everything was understood, I would be out of work. It is rather ironic that the job of a scientist is to understand nature, and if the scientist completely succeeds, the reward is unemployment. But of the many things that concern me in the day-to-day existence of a scientist, waking up one morning and discovering that there are no problems to solve is rather low on the list.

even earlier times goes on by trying to learn how matter behaved in the early universe.

Every day in a handful of particle accelerators throughout the world scientists accelerate protons or electrons to tremendous energies and collide them. In these collisions it is possible to recreate for a brief instant the conditions that have not existed in the universe for fourteen billion years. By studying what happened in the collisions we can gain insight into the origin of the universe in the same way nuclear physics experiments provide information about the synthesis of the elements.

The highest-energy collisions we can study in the laboratory have an energy equivalent to a temperature of 3,000,000,000,000,000° Celsius. The big-bang model predicts that this was the temperature of the universe about 0.000000000004 seconds after the bang. If the early universe resembled the collisions of particles at high energies, then the universe consisted of a primordial soup containing the entire zoo of elementary particles produced in the collisions. By studying the properties of the primordial soup, perhaps we will learn some answers to fundamental questions about the universe.

In the debris of high-energy collisions are the ingredients of the primordial soup. A picture of a high-energy collision is like a snapshot of the primordial soup.

What were the ingredients in the primordial soup?

But for temperatures larger than can be produced in laboratories, our "direct" knowledge of the properties of matter ends, and we can only speculate about the universe when it was hotter than we can study. Acting as our guide in the extrapolation to earlier times is an equation that is one of the cornerstones of modern physics—the field equation of Einstein's theory of gravity. Because the equation is so beautiful, I can't resist writing it:

$$R_{\mu\upsilon} - \frac{1}{2}g_{\mu\upsilon}\mathcal{R} = 8\pi G T_{\mu\upsilon}.$$

Although it is a very technical equation, even beyond the grasp of many physicists, there is one aspect of it anyone can appreciate. The equation has something in common with every other equation, even one as simple as 2 + 2 = 4: Einstein's equation has a left-hand side and a right-hand side.

In the clever way I have written Einstein's equation, which coincidently is also the way Einstein wrote it, the left-hand side describes geometry (the curvature of space and space-time, the expansion of space, and so on), while the right-hand side encompasses the physics of the primordial soup (the fundamental forces and particles). If we want to understand the expansion of the early universe, the left-hand side, then we must have an understanding of the primordial soup, the right-hand side.

Luckily, our understanding of the laws of high-energy physics is sufficiently advanced to allow us to speculate with some confidence about what the universe should have been like at temperatures even higher than can directly be reproduced. Although such a theoretical speculation about the universe might prove unwarranted (that is, wrong), one day it will be tested.

The highest energy we can produce today is tantalizingly close to the temperature corresponding to what we think should be an important transition in the history of the universe. About 10^{-13} seconds into the bang when the temperature of the universe was about 10^{16} degrees Celsius, the fundamental forces of physics are predicted to have more symmetry than they do today.

All the physical phenomena we see around us, from the sun shining to a motion of an electron in a magnetic field, in principle can be explained by the action of just four fundamental forces: the gravitational force, the electromagnetic force, the strong nuclear force, and the weak nuclear force. We are familiar with the gravitational and electromagnetic forces in our everyday experience. The strong nuclear force is also crucial in our everyday life because it is responsible for binding neutrons and protons into nuclei, and quarks and gluons into neutrons and protons. The less familiar weak nuclear force leads to a type of radioactive decay known as beta decay. Although not as familiar, it is nevertheless crucial to our existence. It is the weak reaction that changes a proton into a neutron, which is a step in the generation of energy in the Sun when a helium nucleus (two neutrons and two protons) is made from four protons. Without the weak force, the Sun would not shine.

Just as cosmologists are seeking deeper understanding of the structure of the universe, high-energy physicists are searching for connections between the forces. Past connections between seemingly unrelated forces have led to great revolutions in physics. Newton's realization that the force responsible for apples falling from trees is the same force that regulates the orbit of the Moon was a revolutionary discovery. The nineteenth-century discovery of British scientists James Clerk Maxwell and Michael Faraday that the electric force and the magnetic force, previously thought to be independent phenomena, were really different manifestations of a single electromagnetic force led to sweeping changes in physics—and changed society forever.

In the late 1960s the American physicists Sheldon Glashow and Steven Weinberg and the Pakistani physicist Abdus Salam realized that the weak force and the electromagnetic force were also connected. This electroweak unification is different in character from electromagnetic unification because it is only exact at energies corresponding to temperatures in excess of 10^{16} degrees Celsius.

This electroweak unification energy is just out of reach of current accelerators, but it is close enough to justify extrapolation of what we know to conditions in the universe at the electroweak unification temperature. If electroweak unification is correct (and the Royal Swedish Academy of Sciences thought the evidence was strong enough to award Glashow, Salam, and Weinberg the 1980 Nobel Prize), then there should be a transition to electroweak unity 10^{-13} seconds after the bang. Perhaps the existence of matter can be traced to events that occurred at this time, and we are near the threshold of a deeper understanding of another cosmic mystery.

Further extrapolation in the same spirit suggests that the strong force should be connected to the unified electromagnetic–weak force. If this occurs it would be at a tremendous energy scale, corresponding to a temperature of about 10^{28} degrees Celsius. If the forces are unified in this way, the universe should have undergone another transition 10^{-35} seconds after the bang. Events that occurred at this time might have left an imprint on the universe in the form of the small lumps in the distribution of matter in the early universe.

Although this scenario is speculation at present, experimental evidence may soon be forthcoming. Soon (but not soon enough) we will have accelerators capable of producing the temperatures of weak unification. Strong unification is far, far beyond the reach of even the most futuristic accelerators dreamt of by physicists, but there might be other signals of this unification, such as the decay of isolated protons. One day we will know, and ride the coattails of Einstein's famous equation to earlier times than he ever imagined.

But we know the ride must end sometime. No matter how beautiful or powerful, there is a limit of validity of every theory. We expect Einstein's theory of gravity to break down at energies corresponding to 10^{32} degrees Celsius, or 10^{-43} seconds after the bang. Earlier than this epoch the quantum nature of space and time, which is not accounted for in Einstein's equation, must become important.

Whether or not unification turns out to be correct, the key to the early universe is understanding the fundamental forces and particles. So when flying over the Fermilab ring I see not only an accelerator but also a time machine. While particle physicists are trying to untangle the complexities of the fundamental forces, at the same time they are pushing back the frontier of our understanding of the early universe.

AND EVER WILL BE

After wandering the corridors of the science museum thinking about ancient astronomers, I can't help but wonder how future generations will regard our own efforts to model the universe. Perhaps our grandchildren will take their grandchildren to the museum one rainy Sunday afternoon in the year 2085 and stop in front of a case housing a display of our astronomical tools and cosmological models. It is easy to imagine them shaking their heads in amazement as they behold a computer from the 1990s. The grandfather might reminisce (as grandfathers like to do) that when he was a child his grandfather used to tell him that people often spent long hours squinting at those two-dimensional screens, and actually input letters and numbers into those clumsy-looking pieces of molded plastic by physically pushing keys with their fingers. After reflecting on the inefficient, primitive means we interfaced with silicon, in all innocence the youngster might question why the letters on the keyboard are arranged in such a strange manner, and innocently ask, "Grandpa, what exactly does q-w-e-r-t-y spell anyway?" Undoubtedly, the tools of modern science, digital computers, and other proud triumphs of our technology will one day seem as primitive as the astrolabe analog computer seems to us today.

Somewhere down the hall from the computer might be a display of the big-bang model and the history of the universe as it was understood at the end of the twentieth century. No doubt people will look on our charts and pictures with the same amusement with which we look on the maps of the world made shortly after the discovery of the Americas.

They might see that we have the basic outlines of the universe right, but in our blindness we have omitted the equivalents of entire continents or didn't comprehend the true extent of what we do know.

If the curators of the museum are kind, there will be a small part of the exhibit dedicated to the explanation that just as the generations and generations before us, we looked up to the sky with blind eyes. It should say that just as pioneers on this small planet explored lands once thought to be beyond reach and astronomers extended our understanding beyond our solar system, twentieth-century astronomers struggled to understand regions of the universe in space and time once thought to be beyond the realm of human comprehension. Whether our cosmological view of the universe is right or wrong, or just incomplete, we were brave enough to confront our ignorance and look. We looked with all our might, and with boldness and imagination managed to see a little bit farther than our predecessors. We were not proud of our blindness, but neither were we ashamed of it or intimidated by it, for we chose to look for the light of truth fully cognizant of our blindness.

ELEVEN

The Raw Edge

More than four centuries have passed since that cold, clear November night in 1572 when the young astronomer Tycho Brahe discovered a new star in the sky over Denmark. After a year of studying the new star, he came to the realization that not only do changes occur in the world around us as the result of human activity, but changes also occur in regions of the universe far beyond the reach of our corrupting influence. The appearance of the new star of 1572 did not herald the birth of a star just beyond the orbit of Saturn as Tycho suspected, but it signaled the death of a star twenty thousand light-years away in a **supernova** explosion. Regardless of whether he witnessed a stellar birth or a stellar death, Tycho knew it was a transitory event in a universe that previously had been thought to be unalterable. Although the significance of Tycho's observations may not have been appreciated by his contemporaries, his observations forever put to rest the idea of an unchanging universe.

In the intervening centuries we slowly developed the view that change in the universe is even more ubiquitous than Tycho could have imagined. We now know that stars are born and die in our galaxy all the time. Stellar births are a long, slow, uneventful process, and the death of a star is usually a quiescent event—like old soldiers, they usually just fade away unnoticed into the dark night. Only once every century or so does a star in our galaxy end its life with a large enough bang to be observed from Earth. Because supernovae and other striking changes in the sky are such rare occurrences, one can understand why it was once believed that the heavens are fixed. It is harder to understand why people refused to let go of the belief in the face of Tycho's evidence to the contrary.

Gazing at the sky before the invention of the telescope, Tycho believed that he could behold the entire universe, and that nothing lay hidden from his insightful vision. Tycho could not foresee that less than a decade after his death an Italian astronomer would look at the night sky through a pair of crude lenses and revolutionize astronomy. Galileo's optic tube revealed a multitude of uncharted stars and other previously unknown phenomena in the sky that were invisible to unaided eyes, even eyes as sharp as Tycho's. In turn, little did Galileo know that some of the fuzzy, nebulous objects he saw with his small telescopes were distant "island universes," equal in rank to our own Milky Way galaxy. Even two centuries after Galileo, Herschel did not realize that the nebulae he charted and studied so conscientiously were mostly external galaxies. It would be another century after Herschel before Hubble would unravel the true nature of the spiral nebulae, and even longer before he would observe that distant galaxies were receding from us in a universal expansion. Of course Hubble could not have foreseen at the time of his observations that today we can use his discovery of the expansion of the galaxies to build a model for the origin and evolution of the universe and trace the history of the evolving, changing, expanding universe back nearly fourteen billion years until the time of the first second of expansion.

A time span of four centuries is enormous by any human measure, but it is an insignificant amount of time compared with the age of the universe, or even next to the age of our own little planet. Our understanding of the universe has changed more in that brief time, however, than in the many thousands of years prior to Tycho's sighting of the new star.

The progression in our view of the universe has not been characterized by a slow monotonic increase in knowledge, but our understanding has grown in spurts in a series of small steps, giant leaps, and false starts. When advances have occurred, they have not come easily. In fact, when looking back over the history of cosmology, I am struck by the one common attribute shared by all who have looked at the sky: confusion at the edge of knowledge. No astronomer, not even Kepler or Newton, ever grasped the true enormity and diversity of the universe. It is folly to believe we do now.

Astronomers both ancient and modern believe that the universe is comprehensible, and that the answers to all our questions are written on the sky. But they are all painfully aware that the task of reading the answers is not an easy one. Every cosmologist has spent at least 99 percent of his or her life in confusion, and only the very lucky ones have been able to see something a little more clearly than others, and, by doing so, to remove a bit of our collective blindness. It is as if the universe is just sitting there, patiently waiting for our wits to grow sharp enough to comprehend it. Perhaps this should not be surprising, because after all the universe was around for billions of years before our species evolved the intelligence and curiosity to wonder about it, and it will be there for billions of years after we cease to be. If there is anything the universe has plenty of, it is time and space.

Despite the enormity of the task faced by cosmologists, they are an arrogant lot. It takes a lot of confidence to imagine that one can understand something new about the universe. But in reflective moments, all cosmologists know that the same thing that happened to Tycho, Galileo, Herschel, and all the others will also happen to them: working at the edge of knowledge, they will unknowingly see things in the sky that are important parts of the puzzle, but they will not comprehend their significance.

To try to ferret out the important and interesting objects from the multitude of things in the sky, every cosmologist looks at the universe through a filter of a model, for without the conceptual framework of a model the staggering number of things in the universe would overwhelm anyone. For this reason, most of the work in cosmology today involves building on the framework of the existing big-bang model for the origin of the universe.

Although the big-bang model as we understand it does not provide all the answers, it does seem to be part of the answer. It makes some remarkable predictions, such as the expansion of the universe, the background of ancient photons, and the relative abundance of hydrogen and helium. But we do not use the big-bang model simply because its predictions are in agreement with observations.

One of the oldest reasons to construct a model of the universe is to predict the motions of the planets. The motivation for the predictions varied from culture to culture. The ancient Mesopotamians chiseled into stone tablets that "when Venus is high in the sky, copulation will be pleasurable...," so it is fair to guess that there was quite a demand for astronomers to be able to predict the motions of Venus. But modern science does not develop models solely for such purposes. We don't want a model simply to calculate things or predict things, but to understand nature. We develop models to find a deeper understanding of presently unexplained phenomena. Sometimes even scientists need to be reminded of that.

On the occasion of a solar eclipse in 1994, I was asked to lecture about eclipses to a science class in a local public school. Since I am too lazy to draw my own figures, I copied a figure from an encyclopedia onto an overhead transparency in order to illustrate the Sun–Earth–Moon alignment during a solar eclipse. As soon as I projected the figure on the screen and began the explanation, someone in the back of the

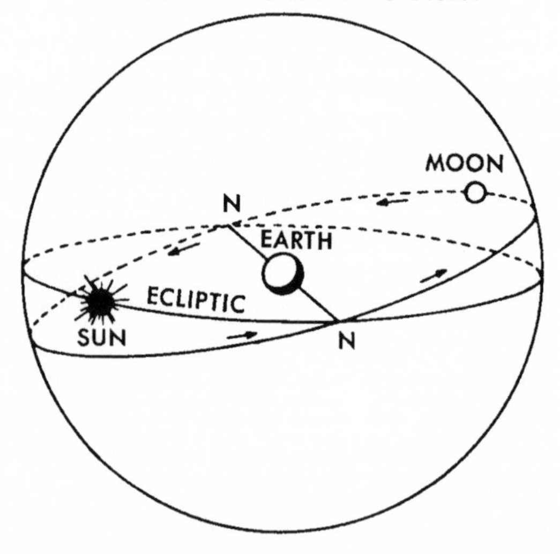

To understand solar and lunar eclipses, one can conveniently work with a solar system model with Earth at the center. This diagram was taken from the nice discussion in the Encyclopedia Americana, *but similar figures can be found accompanying most explanations of solar eclipses. (The letter N represents nodes of the Moon's orbit when Earth, the Moon, and the Sun are all in the same plane.)*

room raised her hand and, in a small voice said, "But Dr. Kolb, Earth isn't at the center of the solar system." The student had noticed something that had escaped my attention. The figure I projected represented a geocentric model for the solar system, with the Sun and Moon orbiting about a stationary Earth.

I used the question as an opportunity to explain that one could predict when eclipses would occur either by assuming a model where the Sun goes around Earth or by assuming a model where Earth goes around the Sun. If the sole purpose of the exercise is to predict when eclipses will occur, it really doesn't matter which model is used. In fact, anyone who actually tried to perform the calculation would be crazy to work with a reference frame of a moving Earth because the mathematics is much easier in the stationary-Earth frame. Of course, that doesn't mean that the person calculating the occurrence of eclipses has to believe that Earth is stationary. In this case the stationary Earth is just a model to be used for calculational ease. You don't have to believe a model is true to use it.

For some reason I thought the class might want to know that there is a fundamental difference between employing a model for calculational convenience in a particular problem, such as using a stationary-Earth model in the calculation of the period of solar eclipses, and using a model to find a deeper understanding of nature. It is the model that leads to deeper understanding and is incorporated into future models that is the truer one. The model that points the way to further knowledge is the model that can be "believed." A model used solely for calculations is nothing more than a means to save the phenomena.

If you want to know whether you should take a model seriously or just regard it as a calculational tool, you can imagine that you are building a wall out of stones. For instance, if the stones represent models for various phenomena seen in the sky, and the wall represents all of astronomy, then choosing a model to explain a phenomenon is like choosing a stone to be incorporated into the wall. Sometimes the stone seems to fit naturally into a space in the wall; more often it has to be trimmed a bit to fit in. But it is impossible to judge whether it is a "beautiful" stone or the "correct" stone for that place in the wall when it is first inserted, because the true beauty or utility of the stone can't be judged in isolation from the rest of the wall. The two real criteria to judge the stone are whether it is one on which other stones can be placed and whether it exists harmoniously with the surrounding stones. If the stone not only fulfills the function of taking up space in the wall but also provides a platform on which to place other stones, it a beautiful stone.

If one imagines the Copernican and Ptolemaic models as foundation stones, in isolation the Ptolemaic model might do as well as the Copernican model for specific calculations, but it is impossible to build on it the planetary laws of Kepler, Newton's laws of motion, or Einstein's theory of gravity. Sometimes future developments are required before it is possible to sort out which model is truer.

Choosing the model is only the start of the process of discovery, however. So I went on to tell the class that even when working within the boundaries of a prescribed model, there is still an art to picking out a sensible problem on which to work, because confusion always lurks at the edge of knowledge.

By its very nature the edge of knowledge is at the same time the edge of ignorance. Viewed from a safe distance from the knowledge side, the edge looks clean and smooth, but when examined up close from the ignorance side, the edge seems dirty, ragged, and raw. Many who have visited the raw edge of ignorance have been cut and bloodied by the experience. I told the class that every scientist visits the raw edge of ignorance, and that even the greatest cosmologists went in the wrong direction or worked on nonproductive problems. Einstein knew the right problem to work on, applying **general relativity** to cosmology, but went off in the wrong direction. Of course, sometimes even the wrong questions lead to spectacular results, as when Johannes Kepler was led to the development of his three laws of planetary motion while trying to answer the question "Why are there six planets?"—a question now considered irrelevant.

I don't know how much of the long-winded exposition was appreciated by the fifth-grade class, but it was the last question I was asked that day.

✦ ✦ ✦

One of the most popular titles for cosmology meetings contain phrases like "Outstanding Problems in Cosmology" or "Future Directions in Cosmology," and there are heated debates at meals and over coffee about the most promising problems on which to work. After all, everyone wants to find the questions whose answers will lead to a deeper understanding of the universe.

If you happen to ask a cosmologist, or any scientist for that matter, what he or she thinks is the most interesting problem in science, more likely than not the answer will be the problem that scientist happens to be working on at that moment. If that is not the answer, find another scientist with whom to talk.

All competent cosmologists can rattle off a list of what they think are the ten most pressing questions about the universe. Although today there would be a fair degree of unanimity in the list, more likely than not most of the questions on the list will turn out to be irrelevant or unfruitful.

It is surprisingly easy to find fundamental questions in cosmology, because in many ways cosmology is a simple science. Hidden beneath a thin (and sometimes not so thin) veneer of mathematical complexity are simple, basic questions. Most of today's most interesting cosmological questions are not even original, but they are the same questions that have been asked ever since we first turned our gaze to the skies. Not only can most of the questions be understood by anyone, but they can be asked by people who have no scientific background but do have the only irreducible requirement for a scientist: curiosity about the universe.

VOX POPULI

Although cosmologists are an individualistic lot, many share a common trait: they are selfish. They find the universe so interesting that they are often unwilling to do anything that will take time away from work. Unfortunately, I am not selfish—or at least not as selfish as I should be, because I have a terrible weakness. In addition to doing it, I like to talk about cosmology. In a typical year I speak to audiences ranging from third graders to fellow cosmologists. Audiences might include only a few dozen people at Kiwanis and Rotary meetings or several hundred people at large public events. Lectures are given in my own zip code as well as in other countries.

I don't know why I can't seem to say no to speaking invitations. Perhaps one of the reasons I like to lecture so much is that I always enjoy answering the questions. It amazes me that people from different backgrounds are sufficiently interested in cosmology to stay awake through the lecture and curious enough about the subject I care so much about to ask questions at the end.

After perhaps a hundred sessions fielding questions from nonscientists, I have noticed that some of the same questions are asked over and over again. In fact, I can assemble what I believe to be the top-ten list of the most asked questions about the universe:

10. Where is the center of the universe?
9. What is beyond the "edge" of the universe?
8. Into what does the universe expand?
7. Do we live in a special place in the universe?
6. Why was there a bang?
5. What happened before the big bang?
4. How old is the universe?
3. How large is the universe?
2. Will the universe expand forever or ultimately recollapse?
1. Of what is the universe made?

All of them are good questions. After all, at one time or another they have all been asked by cosmologists; some of them are still asked. A few of the questions are easy to answer; some can't be answered. Many of the questions are also on the top-ten list prepared by cosmologists.

Although complete answers to the ten questions would fill a book, at least some of the questions have already come up in the course of this book. For instance, questions 8 through 10 are all related to the geography of the universe. Once it is understood that the big bang is not an expansion of galaxies into empty space, as discussed at the end of Chapter 8, it can be appreciated that there is no center and no edge to the universe in the big-bang model. The universe doesn't expand "into" anything because there is no "edge" to the universe.

The answer to question 7 is no, we do not live in a special place in the universe. But perhaps the question would be better phrased as "Are we special?" The answer to that question may well be yes, in light of the fact that in an absolutely boring location in the universe we developed the curiosity to ask the rest of the questions on the list.

The answers to questions 5 and 6 are probably related to events that occurred around the quantum-gravity epoch mentioned in Chapter 10. There is a fairly large effort today to answer these questions, and some of the answers that seem to be emerging are quite exciting. One possible answer to question 6 is that the bang was inevitable, that nothing (a universe without matter, radiation, space, or time) is unstable, and it is possible to imagine the emergence of a universe from nothing.[1] A likely answer to question 5 is that there was no "before" before the big bang, because time itself was created along with everything else in the bang. If those answers seem vague, it is because we simply don't yet know the answers, but we can still make reasonable suppositions based on the little we do understand. It is not known when the answers will be found, or even if they can be found within the framework of the big-bang model or if the model will require extension.

The answer to question 4—"How old is the universe?"—is inextricably tied up with the question of the present expansion rate of the universe, and the expansion rate of the universe is confused by our inability to find reliable indicators of the distance to very distant galaxies as discussed in Chapter 6. Until the issue of the extragalactic distance scale is settled, the best we can do is allow the generous range of between ten and twenty billion years for the age of the universe.

Question 3 is hard to answer. We just don't know how large the universe is. It may be infinite in spatial extent, or space may be curved and close up into a finite volume. To answer the question, we have to measure the curvature of space. It is necessary to travel hundreds of miles on the surface of Earth before the curvature of Earth has to be taken into account. We don't know for sure how far we have to look before the spatial curvature of the universe becomes important. We only know that it is more than thousands of millions of light-years.

Most of the work of a cosmologist involves studying the past. But predictions about the future are also possible. Unfortunately, most of the predictions are about things that will occur billions of years in the future and aren't easily tested. One such prediction is the ultimate fate of the universe, question 2. We know that today the universe is expanding. We also know that there is matter in the universe opposing the expansion. The question of whether or not the universe will expand forever is related to the question of whether the universe today is expanding fast enough to overcome the gravitational pull of everything in the universe trying to pull it back together. In principle, the question could be answered if we could determine the exact values of two fundamental parameters of our universe: its expansion rate and the density of matter. Although the big-bang model provides the framework to answer the question, at present it is impossible to determine the two necessary

[1] In this context, "nothing" means a lot. Not only does it mean a universe without matter or energy, but it also implies a universe without space or time.

parameters with enough precision. As just mentioned, the expansion rate of the universe is hard to pin down. Furthermore, the mass density of the universe is hard to determine because we don't know the answer to question 1, "Of what is the universe made?"

Although question 1 is simple enough to be asked by a child (and it often is), it is a deceptively difficult question. One good answer might be that the universe is made of normal matter, the stuff of stars and planets. But since there are more than a billion ancient photons for every neutron or proton in the universe, leaving them out would not be a good idea. Even including the ancient photons along with normal matter, the answer still seems incomplete. The reason most cosmologists list this question as number 1 on their list of questions about the universe is that almost all of the mass of the universe seems to be invisible.

Cosmologists are usually interested in the big picture and not overconcerned with exactitude. A cosmologist will report that the universe we see is made of hydrogen and helium and ignore the small amount of the other elements so crucial for life. Since 99 percent of the universe is made of hydrogen and helium, a cosmologist needn't be concerned about the rest of the hundred or so chemical elements. Numerical accuracy is usually not demanded of a cosmologist. For "cosmological" accuracy, π is equal to 3; there is usually no need to carry around a lot of numbers after the decimal point. But it is a matter of concern when an accounting is done and the numbers don't add up. Such an embarrassment arose when it was discovered that there is some type of matter out there that hasn't been accounted for.

To find the mass of something is not that difficult, even if you can't see it or put it on a scale. It is a straightforward procedure to determine the mass of the Sun just by measuring the orbital speed of Earth about the Sun and the distance between Earth and the Sun. The same application of Newton's laws of motion can be used to find the mass of our galaxy if we are able to measure the orbital velocity of stars on the fringe of the galaxy and determine how far they are from the center of the galaxy. We can then compare the mass determined from measuring the velocities of stars with the mass we can account for in the form of stars, dust, and other things we see in our galaxy.

When the accounting is done, there is a shortfall that just can't be ignored. We find that the galaxy has a much larger mass than the sum of all the stars, dust, and other things we "see." The shortfall is not just a few percentage points, but most of the mass of our galaxy seems to have been left unaccounted. We know the mass is there because of the gravitational force it exerts on the stars orbiting the galaxy. But it is as if the mass were "invisible."

The problem is not just one of invisible mass in our galaxy, but invisible mass seems to be the major constituent in every galaxy we measure. They all seem to be dominated by invisible matter. If we do the same mass audit for larger objects, say, groups of galaxies, the problem seems to get worse. In fact, most of the mass of the universe, as much as 90 percent, seems to be invisible.

There are many theories about the nature of this ubiquitous, invisible matter. Some of the less exotic solutions involve normal matter that formed objects we haven't yet detected because they don't emit much radiation: giant snowballs, small rocky planets, giant gassy planets, mass-disadvantaged stars of color (small, dim, red-or brown-dwarf stars less than about 10 percent of the mass of the Sun), or even black holes. But many cosmologists believe that the invisible matter might be a species of elementary particles from the primordial soup of the big bang. It could be a type of particle we already know something about, or it could be a yet undiscovered elementary particle.

Whatever the nature of the invisible matter, the search for its identity consumes the efforts of a fair fraction of cosmologists. It is certainly an interesting problem. Only time will tell if it will prove to be a fruitful question.

✦ ✦ ✦

Although the answer to question 1 is unknown, I think I have an idea about what the answer might be. I know that the answer to question 1 on the list of *the* most fundamental issues in cosmology today will likely be something different from what I imagine. Perhaps the answer to the question is staring us in the face but we don't recognize it. I also suspect that years from now people might well wonder why we made such a fuss about it. But perhaps it just might turn out to be a fundamental question, and the answer might be at hand if only we would look a little harder.

If there is any lesson from the story of the people and ideas that shaped our view of the universe, it is that knowledge about the universe is not easily won. In spite of tremendous effort, almost all of the time we fail to find an answer. Most of the time we even fail to ask the right question. But occasionally, through some combination of serendipity, work, genius, and insight, we do see the sky a little more clearly. It is the hope of seeing something no one else has seen, or understanding something no one has understood before that keep astronomers awake so many nights.

Even though I know the probability of success is slim, the possibility of deducing the nature of invisible matter is too tantalizing to pass up. So it is now time for me to end this story and return to the most frustrating and wonderful undertaking I can imagine: working at the raw edge of ignorance as a blind watcher of the sky.

The Devil in the Details

ACTION AT A DISTANCE: Most of us are comfortable with the concept of forces pushing or pulling objects around. But in our everyday experience, the object being pushed or pulled appears to be physically connected to the agent of change. Newton was greatly troubled when he realized that the gravitational force seemed to act across empty space, without the presence of an intervening medium to transmit the force. The phenomenon of one object exerting a force on another object without the intervention of anything physical is called action at a distance. We now know that all forces act at a distance.

ANGULAR SIZE: The size of the opening between two lines that meet at a point can be described in terms of an angle. The most familiar unit for measuring the size of an angle is the degree, with 360° in a full circle. A degree is further subdivided into 60 minutes and each minute is further divided into 60 seconds.

Astronomers describe the apparent size of celestial objects in terms of their angular size, or angular diameter. If two lines are drawn from Earth to opposite edges of the Moon, the angle formed by the two lines at Earth would be about a half of a degree. Another way of saying this is that the Moon subtends an angle of one-half degree, or has an angular diameter of one-half degree.

The angular size can be converted into a true physical size of an object only if the distance to the object is known. Of course, for a fixed physical size, the more distant an object, the smaller its angular size will be.

ARISTARCHUS'S DETERMINATION OF THE EARTH–SUN DISTANCE: Except during a lunar eclipse, the Sun always illuminates the side of the Moon facing it. The Sun–Moon–Earth angle is 180° for new moons and 0° for full moons. Quarter-moons occur when the Sun–Moon–Earth angle is exactly 90°.

As shown in the upper half of the illustration, if the Sun is infinitely far away, then the new moon, the full moon, and the quarter-moons would be equally spaced on its orbit (assuming the orbit is circular and the Moon orbits Earth with constant speed, as believed by Aristarchus).

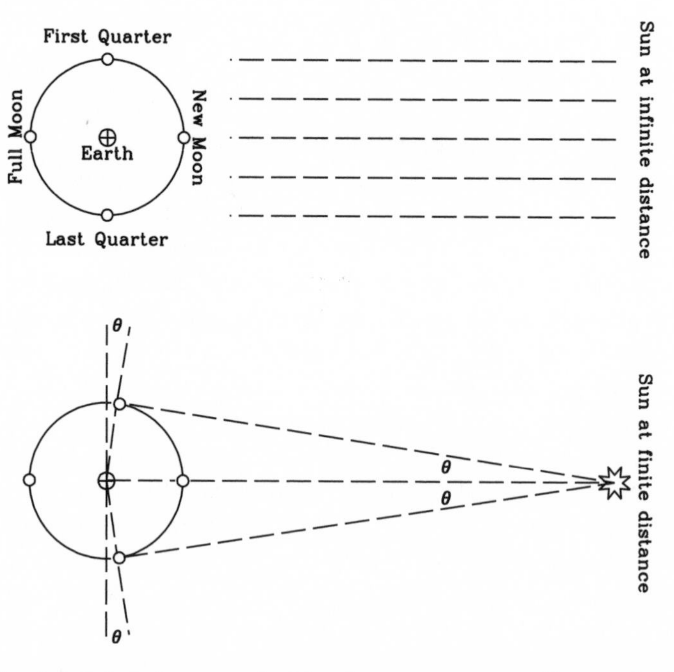

If the Sun is infinitely far away, the Moon's quarters are equally spaced, while if the Sun is a finite distance away, the time between the quarter-moons and the new moon is less than the time between the quarter-moons and the full-moon.

Since the Sun is not infinitely far away from the Moon, the quarter-moons are closer to the new moon than to the full moon (see the bottom half of the illustration). Again assuming that the orbit is circular and the moon travels with constant speed, if the time interval between the new moon and the quarter-moons is subtracted from the time interval between the full moon and the quarter-moons, the measure of the angle θ in the illustration can be determined. This angle can be used to express the Earth–Sun distance in terms of the Earth–Moon distance (which again is known in terms of the radius of Earth, which was learned from the hole in the ground).

Although the reasoning and the geometry used by Aristarchus was flawless, he faced a daunting task, because the time interval between the new moon and the quarter-moons is only a half hour shorter than the time interval between the full moon and the quarter-moons. Without reliable clocks he couldn't measure a half-hour difference over the twenty-eight-day orbital period of the Moon, and he thought the difference was twelve hours, rather than a half hour. This led him to believe that the angles were much larger than they actually are, and led to an underestimate of the distance to the Sun by a factor of about twenty.

THE CELESTIAL SPHERE: It was once thought that Earth was the center of the universe, and the stars were attached to the surface of a sphere surrounding Earth. This sphere was known as the celestial sphere. Of course, the stars are located at various distances from us, and Earth is not the center of the universe, but the concept of the celestial sphere is still a useful way to locate the positions of objects on the sky as viewed from Earth.

If we project Earth's equator out into space, we obtain a plane known as the celestial equator. Projecting Earth's north and south poles into space defines the north and south celestial poles. The location of any object on the celestial sphere can be specified by angles with respect to the celestial equator and poles.

Suppose you go out one night and discover a new star. Since you discovered it, you get to name it, and you decide to call it *Bob.* Of course you want others to look at Bob, but you have to tell them where to look. Since the sky is a big place with lots of stars, you just can't say, "Look up there somewhere"; you must specify Bob's coordinates. Since space is three dimensional, in principle you must give three numbers to specify uniquely the position of Bob. But one of the numbers could be taken to be Bob's distance from Earth. If you want someone just to see Bob, but not to visit him, it is not necessary to know the distance, so the position of Bob in the sky can then be given in terms of two numbers. These two numbers can be chosen to be Bob's position on the celestial sphere, usually specified in terms of two angles known as the declination and the right ascension. These angles are equivalent to the angles of latitude

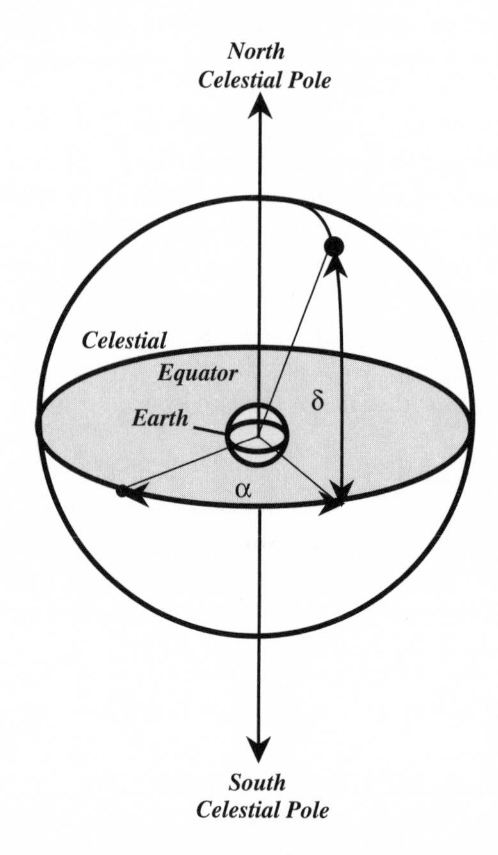

The equator of the celestial sphere is found by projecting Earth's equator into space, and the poles of the celestial sphere result from extending the line connecting Earth's poles. The position of any object on the celestial sphere can be given in terms of two angles, its right ascension denoted by α *and its declination denoted by* δ*. Of course, fixing the right ascension and declination fixes only two of the three coordinates. The distance of the object from Earth is not specified, nor is that knowledge necessary to find an object in the sky.*

and longitude used to specify the position of any point on the surface of Earth. On Earth, latitude is measured as degrees north or south of the terrestrial equator, whereas on the celestial sphere, declination is measured in degrees north or south of the celestial equator. On Earth, longitude is given as an angle east or west of the terrestrial prime meridian in Greenwich, England, whereas on the celestial sphere right ascension is defined as the angle from the celestial "prime meridian." This celestial prime meridian is defined to be the point on the sky known as the vernal equinox, where the Sun crosses the celestial equator from south to north marking the beginning of spring in the Northern Hemisphere.

THE COPERNICAN SYSTEM: In 1543, Copernicus published his epic work *On the Revolutions* proposing a heliocentric model of the solar system. The diagram of the Copernican arrangement in Chapter 2 represents every bit as much of a simplification as does the corresponding diagram of the Ptolemaic model. As Swerdlow and Neugebauer put it, "Anyone who thinks that the Copernican theory is 'simpler' than Ptolemaic theory has never looked at Book III of *The Revolutions*." In fact, Copernicus constructed a cosmology employing more circles than Ptolemy.

The Copernican system *as proposed by Copernicus* is neither simpler nor more intuitive, nor does it do a much better job of agreeing with observations. So why, then, is Copernicus a hero and "**epicycles**" the ultimate pejorative description of an ugly scientific model? It is because Copernicus's model was an enormous step in the right direction. Newton could connect astronomy with the sciences of mechanics and dynamics using Copernicus's model, whereas Ptolemy's model was barren. In this respect, Copernicus had the truer model.

DEFERENT: See **THE PTOLEMAIC SYSTEM.**

DYNAMICS: The science of dynamics studies how bodies behave under the action of forces. Several people before Newton, most notably Galileo and Descartes, developed the idea of inertia, which in the context of Newtonian physics means that a body travels in a straight line unless acted on by a force. Newton completed the picture of inertia and extended the study of the motion of bodies to include their behavior when acted on by a force. Forces, momentum, acceleration, and the motion of bodies were all connected in the science of dynamics.

ECCENTRIC: See **THE PTOLEMAIC SYSTEM.**

ECCENTRICITY: See **ELLIPSE.**

EINSTEIN'S EQUATIONS: The one-sentence explanation of Einstein's theory of gravity usually states that matter tells space how to curve, and the curvature of space tells matter how to move. That simple idea is elegantly embodied in the mathematical equation $R_{\mu\nu} - \frac{1}{2}g_{\mu\nu}\mathcal{R} = 8\pi G T_{\mu\nu}$. This equation is known as Einstein's equation, or sometimes Einstein's equations because the equation above is actually a compact notation for ten separate equations. The Ricci curvature tensor $R_{\mu\nu}$, the metric tensor $g_{\mu\nu}$, and the Ricci curvature scalar $\mathcal{R}$ are determined by the curvature of spacetime, while the stress-energy tensor $T_{\mu\nu}$ depends on the distribution of matter. The constant G is just Newton's gravitational constant. A more user-friendly way to write the equation might be CURVATURE OF SPACETIME $= 8\pi\ G \times$ MATTER. Solutions to Einstein's equations are notoriously difficult to find unless the distribution of matter is very simple. An example of a simple system is when matter and radiation are distributed in a uniform way throughout space, as seems approximately to be the case in our universe. Solutions to Einstein's equations then describe a universe that is either expanding or contracting.

ELECTROMAGNETIC SPECTRUM: The complete range of electromagnetic radiation is contained in the electromagnetic spectrum. The wavelength of an electromagnetic wave λ is related to its frequency ν and the speed of light c by $c = \lambda\nu$. The longest wavelength electromagnetic waves are called radio waves, and the shortest wavelength waves are gamma rays. Visible light is nothing more than electromagnetic radiation in a very narrow range of the electromagnetic spectrum.

ELLIPSE: An ellipse looks like a squashed circle, stretched in one direction and flattened in the other direction. But an ellipse obeys a precise mathematical equation; it is not simply a deformed circle. The degree to which an ellipse departs from a circle is defined by its eccentricity. The eccentricity of an ellipse is a determined from the ratio of the length of the long (major) axis of the ellipse to the short (minor) axis of the ellipse. If the lengths are equal, the ellipse is a circle and the eccentricity is zero. If the length of the semimajor axis is indicated by a and the length of the semiminor axis indicated by b, then the eccentricity of

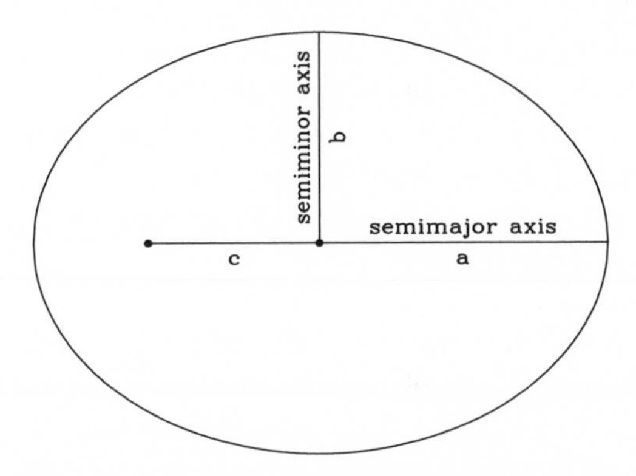

An example of an ellipse with eccentricity 0.6. The semimajor axis, the semiminor axis, and one of the foci indicated. The other focus is on the opposite side of the ellipse on the major axis, the same distance from the center.

the ellipse, *e*, is defined as $e = \sqrt{1 - a^2/b^2}$. For the ellipse pictured above, $e = 0.6$. Just as the circle is the set of all points equidistant from a point known as its center, an ellipse is the set of all points where the sum of distances from two points, called the foci, is the same. The foci are aligned on the major axis of the ellipse, a distance $c = e/a$ on either side of the center of the ellipse.

EPICYCLE: See **THE PTOLEMAIC SYSTEM.**

EQUANT: See **THE PTOLEMAIC SYSTEM.**

GENERAL RELATIVITY: See **RELATIVITY.**

INERTIA: The property that requires a force to act on a body to change its state of motion.

INVERSE SQUARE: "Inverse square" is a quick way to say "proportional to the power −2." For instance, the intensity (or apparent brightness) of a light source depends on the inverse square of the distance. If a light bulb with a particular brightness is moved twice as far away, the apparent brightness will decrease by a factor of $\frac{1}{2}^2$, or $\frac{1}{4}$. If the bulb would have been moved three times as far away, the apparent brightness would have decreased by a factor of $\frac{1}{3}^2$, or $\frac{1}{9}$. The strength of the gravitational force also varies as the inverse square of the distance. On average, Neptune is about thirty times as far from the Sun as Earth, so the gravitational force of the Sun felt by Neptune is a factor of $1/30^2 = 1/900$ times smaller than the force felt by Earth. If the **luminosity** of an object is known and its brightness measured, its distance can be determined by the inverse-square relation APPARENT BRIGHTNESS = LUMINOSITY/(4π × DISTANCE-SQUARED).

KEPLER'S LAWS: Kepler discovered that there are three laws that describe the motion of the planets. The first law states that the planets travel in elliptical orbits about the Sun, with the Sun at one of the foci of the **ellipse**. Kepler's second law relates the orbital velocity of a planet to its position in the orbit, with the planet traveling faster when nearer the Sun. Kepler's third law relates the time it takes for a planet to complete a full orbit to its distance from the Sun. The more distant planets take longer to circumnavigate the Sun. Kepler's laws of planetary motions give an excellent, but not exact, description of the solar system. Small departures from Kepler's laws are caused by the influence of the other planets. Newton realized this and attempted to estimate the size of the effect.

LUMINOSITY: The luminosity of an object is a measure of how much energy per time is emitted. A familiar measure of luminosity is the wattage of a light bulb. A 400-watt bulb has four times the luminosity of a 100-watt bulb; that is, it emits four times as much energy per second. The luminosity is an intrinsic property of an object, so it does not depend on how far away it is. Because of the **inverse-square** law, a 400-watt bulb twice as far away as a 100-watt bulb will *appear* equally bright (that is, they will have the same brightness), but the 400-watt bulb will still have four times the luminosity of the 100-watt bulb.

NUCLEOSYNTHESIS: The process of building nuclei out of other nuclei is known as nucleosynthesis. For instance, a hydrogen nucleus (a proton) can be combined with a neutron to form a heavy isotope of hydrogen known as deuterium. Two deuterium nuclei can combine and form the common isotope of helium.

PARALLAX: The *apparent* displacement of an object when viewed from two different positions is known as parallax (see the illustration on page 34). If the angle of displacement can be measured, and the distance between the points of observations is known, then the distance to the object can be determined. More distant objects have a smaller displacement angle. The word *parallax* refers to the phenomenon as well as to the angle of the displacement.

THE PTOLEMAIC SYSTEM: The diagram of the Ptolemaic arrangement in Chapter 2 is a greatly simplified version of the model. The diagram shows only seven concentric circles, but the complete Ptolemaic model contained more than thirty. Although the simple diagram conforms to the dictum of Plato that heavenly bodies travel in perfect circles,

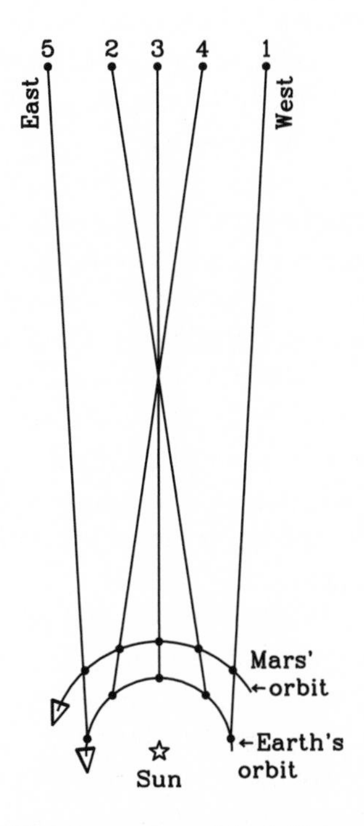

Retrograde motion: The apparent position of Mars as viewed from Earth changes as the planets orbit the Sun. The five points indicated on the drawing represent the positions of the planets separated by roughly forty days. Mars travels more slowly in its orbit than Earth (see **Kepler's Laws***), so Earth, traveling on the inside track, at times "overtakes" Mars and passes it. Between times 1 and 2 Mars appeared to travel eastward, between 2 and 3 and between 3 and 4 Mars appeared to travel westward in retrograde motion, and between 4 and 5 Mars appeared to resume its eastward journey.*

such a simple model cannot account for the motion of the planets; in particular, it cannot explain their retrograde motions.

Planet comes from the Greek word for "wanderer," and indeed the planets seem to wander among the "fixed" stars. In general, the planets seem to move slowly eastward across the constellations of the zodiac. Occasionally, however, the planet will seem to stop its eastward motion and travel westward. This westward movement of the planet is referred to as retrograde motion.

It is easy to illustrate the phenomenon of retrograde motion by considering the orbit of Mars. Comparing the orbit of Mars and the orbit of Earth, Earth has the inside track: it is closer to the Sun and has a larger orbital velocity. Because Earth travels around the Sun more rapidly than Mars (Mars requires 1.88 years to complete an orbit), Earth can overtake and pass Mars. When it does, Mars seems to travel in retrograde motion, just as when one overtakes a slower car on the highway, the slower car appears to travel backward.

Any cosmology with Earth at rest must have some mechanism to explain the retrograde motions of the planets. In the Ptolemaic system it was by the introduction of epicycles. As viewed from Earth, the epicycle travels eastward on something known as the **deferent** (the deferent is literally a circle that carries another circle; the other circle in this case is the epicycle). If the motion of the planet on the epicycle is also in the eastward direction, it will simply add to the general eastward motion. However, if the motion of the planet on the epicycle is in the westward direction, the westward epicyclic motion will subtract from the eastward motion along the eccentric, at times leading to an overall westward motion of the planet.

Although this relatively simple model of one epicycle and one deferent per planet can explain retrograde motion, even more complexity is required to describe accurately the planetary motions. For instance, because the planets do not travel with uniform velocity it was necessary to remove Earth from the center of the deferent and construct an eccentric circle centered on a point outside Earth. The angular velocity of the epicycle traveling on the deferent was constant when viewed from yet another point interior to the deferent, known as the equant. The equant corresponds roughly to the empty focus of the **ellipse** in Kepler's model. In addition, the deferent could be tilted one way in space, the epicycle in the opposite way, and the whole contraption could rock back and forth. The deferent could also perform circular motions, and so on. The purpose here is not to give a

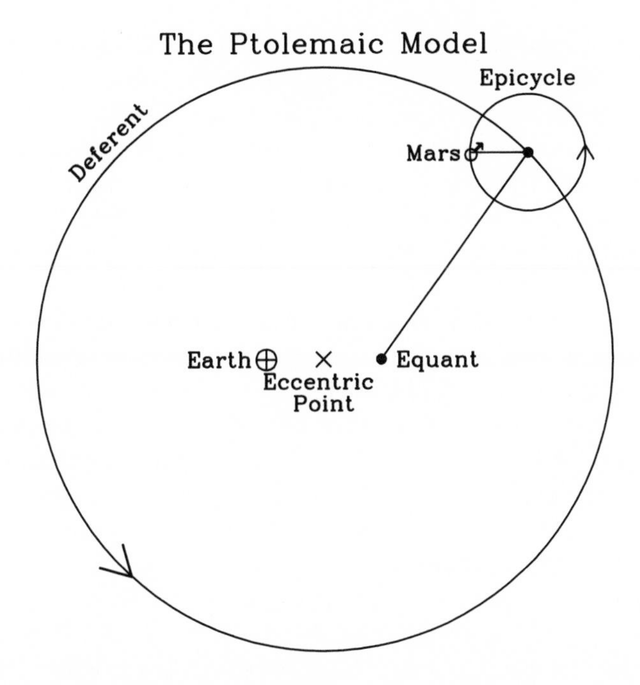

The Ptolemaic model for the orbit of Mars: The planet moves in a circle called the epicycle, *and the center of the epicycle moves on an* eccentric circle *called the* deferent. *The center of the eccentric circle is not Earth, but a point outside Earth. The center of the epicycle moves such that its angular motion about a third point, called the* equant, *is constant.*

description of Ptolemaic cosmology, but to give some feeling for the complexity and beauty of the model.

Although the model was unwieldy and seems baroque by modern standards, it is ingenious and should assume its rightful place as a triumph of ancient science. With enough circles, the model does a remarkably good job of predicting the location of the planets. However, over the span of a millennium, the cumulative effect of small errors between the predictions and the observations made apparent the inadequacy of the model.

QUASARS: Very distant, very luminous objects with a starlike appearance are known as quasars. Quasars are so small and so distant that on photographic plates they resemble stars. Although the energy-producing regions of quasars are small (probably less than a light-day across), their **luminosities** exceed the energy output of our entire galaxy. They are believed to be the superluminous cores of distant galaxies, with the tremendous energy output resulting from the energy released when stars are swallowed by massive black holes.

RELATIVITY: Einstein developed two theories of relativity—the *special theory of relativity* in 1905 and the *general theory of relativity* in 1915. The special theory is a theory of the unaccelerated motion of bodies. Familiar results from the special theory are that velocities are limited by the speed of light, measures of time and distance are relative to the velocities of the observers, and one should think of space and time not as independent but as intimately related. General relativity is Einstein's theory of gravity. It is an extension of his special relativity to accelerated observers. Familiar results from the general theory are that space is curved, that the path of light can be bent by a massive object, and the formation and evolution of black holes and worm holes.

RETROGRADE MOTION: See **THE PTOLEMAIC SYSTEM.**

SPECIAL RELATIVITY: See **RELATIVITY.**

SPECTROSCOPY: When light, or any **electromagnetic radiation**, is emitted by a body, it is usually emitted over a range of wavelengths. The intensity of the radiation as a function of the wavelength is the spectrum of the radiation. Spectroscopy is a branch of science concerned with the analysis of spectra.

STELLAR ABERRATION: The apparent displacement of a celestial body from its actual position because of the motion of Earth is known as aberration. If we look at the streaks of rain on the side window of a moving car, it appears that the rain is coming from some angle rather than overhead. The apparent angle depends on the velocity of the car. The faster the speed, the more tilted away from vertical the rain appears.

If one realizes that the platform from which we observe the stars is moving as Earth orbits the Sun, then one expects that the apparent direction of the light from the stars that rains down on us should be displaced because of the motion of Earth. Since the orbital velocity of Earth about the Sun is only about 0.01 percent of the velocity of light, the apparent displacement is very small, less than about twenty seconds of arc. Although the orbital speed of Earth is very nearly constant, the direction of the velocity changes as Earth goes around the Sun, so the apparent position of the star changes throughout the year.

Although the apparent shift in the position of a star due to aberration is small, it is usually much larger than the shift in the apparent position due to **parallax**, and must be removed before the parallax can be determined. The amount of aberration does *not* depend on the distance to the star, whereas the parallax does.

SUPERNOVA: We are confident that Tycho indeed observed a supernova, because modern radio telescopes can locate a remnant of a stellar explosion at the position on the sky where Tycho reported seeing the new star.

Supernovae are divided into two types, imaginatively known as Type I and Type II. If the spectrum of the light from a supernova shows the presence of large amounts of hydrogen, it is called a Type II supernova, and if it doesn't, it is classified as Type I. Supernovae are further subdivided into subclasses, Ia, Ib, and so on. The number and characteristics of the subdivisions are largely a matter of taste. Type I supernovae are believed to be caused by the explosion of a white-dwarf star, whereas Type II supernovae are thought to represent the explosion that occurs at the end point of the evolution of giant stars more than about eight times the mass of the Sun. The description in Chapter 2 of the presupernova evolution of Tycho's supernova assumed it was Type II.

Of course, we do not know whether hydrogen was present in the spectrum of Tycho's supernova, so we can't say definitively whether it was Type I or Type II. However, the rate at which its brightness decreases in time is another discriminant between the two types of supernovae. From recorded observations of the rate of dimming of the supernova, mostly made by Tycho, it is considered likely that it was a Type I supernova.

Bibliography

This bibliography is a guide for anyone interested in gathering more information about any of the subjects discussed in this book. The references given here represent just a few selections from a reasonably large body of literature. Since the goal is to provide a starting point for further exploration rather than a complete itinerary for a journey, for the most part I mention only those books that could be found easily by people without access to libraries with extensive holdings. Many classic books have been reprinted over the years by different publishers, and usually I refer to the most recent printing. Books that are long out of print are mentioned only if they had a particularly important role in the development of the subject. In the same spirit, I refer to articles in scholarly journals only in the absence of books covering the same material.

Many quotes by historical figures appear in the early chapters of the book. Rather than encumber the bottom of every page of text with lots of *ibid., loc. cit.,* and *op. cit.* citations, I include references to quotations in this section. Furthermore, I include references only to the more significant quotations. The quotations are identified by page number and the first few words of the quote.

Everyone who writes of the historical development of science must seek the right proportion between presenting dry facts and breathing life into the characters they portray. It is difficult to understand the feelings of another person, and even more difficult to identify with someone from another age, so every biographer views the subject from a slightly different perspective. To a greater or lesser degree, every historical account involves some embellishment of historical fact, or at least some interpretation of historical events, which introduces the potential that the breath of life might contaminate the subject with the opinions and views of the author. My own viewpoint is that of a cosmologist who experiences on a daily basis the struggle to understand the universe. A cosmologist does not have to be of Galileo's stature to know what it is like to look up into the dark night sky and try to comprehend the universe. Although the tools and sophistication of astronomy and cosmology have grown in the intervening centuries, the feeling must be the same.

ONE: EYES ON THE SKIES

If this book has kindled an interest in the billions of points of light we see in the night sky, there is no shortage of good introductory textbooks that can provide a more detailed description about modern as well as ancient astronomy. I don't really have a favorite to recommend, but these are some of the more popular ones:

- J. M. Pasachoff, *Astronomy: From the Earth to the Universe* (Saunders, 1991).
- W. J. Kaufman, *Universe* (Freeman, 1994).
- M. W. Friedlander, *Astronomy from Stonehenge to Quasars* (Prentice-Hall, 1985).

On a slightly more technical level, here are two excellent books about astronomy and cosmology:

- F. H. Shu, *The Physical Universe: An Introduction to Astronomy* (University Science Books, 1982).
- D. Layzer, *Constructing the Universe* (Scientific American Books, 1986).

Finally, for the truly scientifically adventurous interested in the details of modern big-bang cosmology, my favorite source is E. W. Kolb and M. S. Turner, *The Early Universe* (Addison-Wesley, 1990).

TWO: SMASHING THE CELESTIAL SPHERES

In Tycho's last years he wrote a small autobiographical sketch of his life, which, as might be expected, does not exactly understate his remarkable achievements. (The words *modesty* and *Tycho* are not often found together.) The first biography of Tycho was by Pierre Gassendi in 1654, who had access to many of the original documents in Denmark. The significance of this biography is that it was written so soon after Tycho's death that many of the details of his life and work that otherwise would have been lost or forgotten have been preserved. In part because of Gassendi's effort, we have a wealth of information about Tycho compared with the sketchy information we have about his predecessor Copernicus, who died just three years before Tycho's birth.

Johan Ludvig Emil Dreyer (1852–1926) was a famous astronomer and historian of science. His 1890 biography of his Danish countryman, titled simply *Tycho Brahe*, has been reprinted several times and can still be found on many library shelves. Dreyer also edited a fifteen-volume collection of Tycho's complete works and correspondence. This massive work, *Tychonis Brahe Dani Opera Omnia*, is a primary source for Tycho historians. All quotes in the chapter may be found there, but unless the reader enjoys translating sixteenth-century Latin, Dreyer's volumes are best left in the hands of scholars.

The best modern biography of Tycho is by Victor E. Thoren, *The Lord of Uraniborg: A Biography of Tycho Brahe* (Cambridge University Press, 1990). This is a well-written, very readable account of Tycho's life and science. A biography of a scientist is empty without some description of the science, which for a nonscientist is often a deterrent to reading them. However, Thoren does a wonderful job presenting the scientific concepts in ways comprehensible to the lay reader. This is the place to start for further reading about Tycho.

We know much less about the life of Copernicus. Much of what we do know can be found in the introduction of the two-volume book *Mathematical Astronomy in Copernicus's De Revolutionibus* (Springer-Verlag, 1984), by Noel M. Swerdlow and Otto Neugebauer. This book is a comprehensive treatment of mathematical astronomy for the serious scholar, but the introduction is accessible to anyone.

Scholarly journals such as *Journal for the History of Astronomy* or proceedings of professional conferences are not readily available at the corner library, but occasionally a useful collection of short articles from journals or proceedings is bound together in a book. A volume of such articles by Owen Gingerich appears in the book *The Eye of Heaven:*

Ptolemy, Copernicus, Kepler (AIP Press, 1992). The articles are not too technical and are enjoyable to read.

Another useful compilation of short articles by leading historians as well as modern cosmologists is *Cosmology: Historical, Literary, Philosophical, Religious, and Scientific Perspectives* (Garland, 1993), edited by Norriss S. Hetherington. The articles in this book are on subjects from the Greeks to modern cosmology.

A brief historical account of Greek astronomy, along with chapters on Copernicus, Brahe, and Kepler, can be found in another book by J. L. E. Dreyer, *A History of Astronomy from Thales to Kepler.* Although originally published in 1905, it has been reprinted several times, most recently as a Dover paperback (Dover, 1953), and it still may be found in bookstores. The very entertaining book *Coming of Age in the Milky Way* (Doubleday, 1989), by Timothy Ferris, also has an excellent treatment of aspects of Greek astronomy. The three-volume set *A History of Ancient Mathematical Astronomy* (Springer Verlag, 1975), by Otto Neugebauer, is highly recommended for anyone interested in an exhaustive study of ancient astronomy.

The most readable discussion of Greek astronomy, as well as Copernicus through Newton, is *The Sleepwalkers,* by Arthur Koestler, readily available as a paperback (Arkana, 1989). Although as a professional cosmologist I often disagree with his unorthodox interpretations or emphasis, I have been greatly influenced by his style. Koestler also seems to be either revered or scorned by professional historians. Love him or hate him, you should read him, for no one else writes on the subject with the flair or wit of Koestler.

The chapter opens with a discussion of the supernova of 1572. A reference for those interested in historical records of this most spectacular of astrophysical phenomena is *The Historical Supernovae* (Pergamon, 1977), by D. H. Clark and F. R. Stephenson.

Dreyer's historical scholarship has been mentioned several times. It is amusing to note that as a professional astronomer he also weaves in and out of the story of the nebulae told in Chapter 7. In 1874 he worked as an assistant to the earl of Rosse at the Parsonstown Observatory, home of the giant seventy-two inch, four-ton telescope known as the Leviathan. It was this telescope that first revealed the important fact that some nebulae have a spiral form. In 1888 Dreyer compiled the *New General Catalogue of Nebulae and Clusters of Stars,* which was based on the surveys of the Herschels and superseded the earlier catalog of

Messier. With supplements added in 1895 and 1908, the *New General Catalogue* (usually abbreviated *NGC*) is still the standard reference catalog, and many astrophysical objects are referred to by their NGC number. For instance, the Andromeda galaxy is known as either M31 or NGC 224, the 31st entry in Messier's catalog and the 224th entry in Dreyer's *New General Catalogue.*

- p. 18: "...in 1006 there was a very great famine..."
Clark and Stephenson, *The Historical Supernovae.*
- p. 19: "*O crassa ingenia. O cœcos cœli spectatores.*"
Dreyer, *Opera Omnia*, p. 12.
- p. 30: "I don't see why the most important problem..."
This is an imaginary quote by a fictional character.
- p. 37: "...[how dare you] not blush to act as if you were my equal..."
The original source is Dreyer, *Opera Omnia.* I have quoted the translation in Thoren, *The Lord of Uraniborg.*
- p. 40: "Tycho accompanied Royal Councillor Minckwiz to supper..."
Written in Kepler's hand in Tycho's logbook. Translation by Edward Rosen, in *Three Imperial Mathematicians: Kepler Trapped Between Tycho Brahe and Ursus* (Abaris Books, 1986).

THREE: WARS OF THE WORLDS

There are a few excellent biographies of Kepler, but the subject is not nearly exhausted. He was such a wonderfully interesting and complicated character, with genius, neurosis, comedy, tragedy, and triumph intertwined throughout a life set against a background of the tumultuous times of the late sixteenth and early seventeenth centuries, that I am amazed that there aren't dozens of movies and books, or at least an opera or two, about his life. Whether the correct place to picture Kepler is standing high on a pedestal or lying down on a psychologist's couch is unclear to me.

Max Caspar's *Johannes Kepler* (Dover, 1993) is the traditional biography, and Koestler's *The Watershed* (University Press of America, 1985) is the unorthodox version. (*The Watershed* is a reprint of the Kepler chapters in Koestler's *The Sleepwalkers.*) For technical aspects of Kepler's work, Bruce Stephenson's *Kepler's Physical Astronomy* (Springer-Verlag,

1987) and *The Music of the Heavens* (Princeton, 1994) should satisfy most everyone. In this last book, Stephenson argues that there is nothing "mystical" about Kepler's feelings and works.

Gesammelte Werke, by Walther von Dyck and Max Caspar, is a complete collection of Kepler's works and voluminous correspondence. All quotes may be found there if the reader cares to pick through the combination of Latin and medieval German.

- p. 47: "At the age of four I nearly died of smallpox..."
- p. 48: "I hated Kolinus ...
- p. 48: "At the age of 21 I was offered union with a woman..."
- p. 49: "That man has in every way a dog-like appearance..."

These quotes were compiled by Kepler in the course of writing the family horoscope. They are translated by Koestler in *The Sleepwalkers,* and the originals can be found in *Gesammelte Werke.*

- p. 52: "...the triangle is the first figure in geometry..."

Kepler describes this in the introduction to *The Cosmic Mystery.*

FOUR: THE GALILEO EQUATION

More has been written about Galileo than about Copernicus, Tycho, and Kepler combined. Although there are many excellent biographies to choose from, the best way to learn of Galileo is to read Galileo. His writings are classics of literature as well as science. Even four hundred years after they were written, they seem fresh and easy to read, for Galileo wrote with a wit and wickedness that has hardly been equaled.

But it seems that not even scientists read Galileo. Although twentieth-century authors and artists still learn to illuminate their characters by reading the words of Shakespeare or studying the canvasses of Caravaggio, very few scientists read the works Galileo, a contemporary of Shakespeare and Caravaggio, to discover how he read the lights in the sky. A good place to start to read Galileo is an abridged collection of Galileo's writings in the book *Discoveries and Opinions of Galileo* (Anchor Books, 1990), by Stillman Drake. Drake's translation of *Dialog Concerning the Two Chief World Systems* (University California Press, 1967) also contains a wonderful foreword by Albert Einstein.

The primary source of Galileo material is the twenty-volume *Le Opere di Galileo Galilei* (Edizione Nazionale, 1964–1966), but of course many secondary sources exist. Two very recent biographies of Galileo are *Galileo: Decisive Innovator* (Blackwell, 1994), by Michael Sharratt, and *Galileo: A Life* (HarperCollins, 1994), by James Reston, Jr. The book *Galileo* (Oxford University Press, 1980), by Stillman Drake, is also highly recommended.

The classic study of the trial of Galileo is by Giorgio de Santillana, *The Trial of Galileo* (Time-Life Books, 1981), but much historical scholarship has been done since this book was first published in 1955. Another source for those interested in the political shenanigans surrounding the trial of Galileo is *Galileo Affair: A Documentary History* (University California Press, 1989), by Maurice A. Finocchiaro.

Those interested in the complicated interplay between Galileo and the church would benefit from *Galileo, Bellarmine, and the Bible* (Notre Dame Press, 1991), by Richard Blackwell, *Galileo: For Copernicanism and for the Church*, by Annibale Fantoli, trans. G.V. Coyne, S.J. (Vatican Observatory Publications, 1996), and *Essays on the Trial of Galileo* (Vatican Observatory, 1989), by Richard Westfall. Bellarmine is a fascinating study in his own right. The most complete biography of him is *Robert Bellarmine, Saint and Scholar* (Newman Press, 1961), by James Brodrick, S.J.

- p. 78: "If he [Grassi] wants me to believe that the..."

- p. 80: "You cannot help it, Signor Sarsi,..."

Galileo, in *Il Saggiatore.*

- p. 91: "For it is the duty of an astrologer [sic]..."

Osiander's preface to *Die Revolutionibus.*

- p. 92: "...the subject which Copernicus is dealing with..."

In the Church's instructions regarding corrections to *Die Revolutionibus.*

- p. 98: "Or another marvels at either the heart of the..."

Maffeo Cardinal Barberini's poem "In Dangerous Adulation."

- p. 101: "...nothing physical which sense-experience sets..."

Galileo, in *Letter to the Grand Duchess Christina.*

- p. 102: "**Sagredo:** [If the Copernican idea is right] why..."

- p. 103: "...that I once heard from a most eminent..."

These two quotes are from Galileo's *Dialog.* I have quoted from the translation of Stillman Drake.

- p. 104: "The constitution of the universe..."

Galileo, in the dedication of the *Dialogues*, as translated by Drake.

- p. 105: "Take note, theologians..."
Note by Galileo in his own copy of the *Dialog.* It was not intended for publication.

- p. 108: "But whereas—after an injunction had been given to..."
This is from the statement Galileo was forced to read at the conclusion of the proceedings against him. The translation is from de Santillana.

- p. 110: "If Galileo had known how to keep himself in favor..."
Grienberger was the successor to Clavius in the astronomy chair of Collegio Romano. Quoted in Westfall, *Essays on the Trial of Galileo.*

- p. 112: "Philosophy is written in this grand book..."
Galileo, in *Il Saggiatore.*

FIVE: NEWTON AT A DISTANCE

Richard Westfall's *Never at Rest: A Biography of Isaac Newton* (Cambridge University Press, 1990) is an exhaustive scientific biography of Newton. The abridged version of this massive work, *The Life of Isaac Newton* (Cambridge University Press, 1993), does a remarkable job of condensing the larger work into a manageable size.

One of the most influential early biographies of Newton is David Brewster's *Memoirs of the Life, Writings and Discoveries of Sir Isaac Newton.* Originally published in 1855, it has been reprinted many times and often appears on library shelves. Nonscientific aspects of Newton's life can be found in *Portrait of Isaac Newton* (Quality Paperbacks, 1990), by Frank Manuel.

Of course, these references only begin to touch on the many books written about Newton. So extensive is the list that there is a book of books of Newton, *Newton and Newtoniana, 1672–1975* (Folkestone, 1977), by Peter and Ruth Wallis. A useful biographical essay about Newton references and sources can also be found in Westfall's book.

- p. 129: "...whilst he [Newton] was musing in a garden..."
A long discussion about this famous quote can be found in Westfall, *Never at Rest: A Biography of Isaac Newton*, p. 154.

- p. 131: "And the same year [1665] I began to think of gravity..."
A discussion of this quote can be found in Westfall, *Never at Rest: ABiography of Isaac Newton*, p. 143.

- p. 134: "It is inconceivable, that inanimate brute matter..." This appears in a letter to Bentley from Newton dated February 25, 1693.

SIX: THE THIRD DIMENSION

The material in this chapter is drawn from a number of sources. The determination of the distance scale is treated in a number of basic astronomy textbooks, and is the subject of a book of its own: *The Cosmological Distance Ladder* (Freeman, 1985), by Michael Rowan-Robinson. The first chapter of Rowan-Robinson's book provides an overview and a history of the distance scale, and is understandable to a wide readership. The subsequent chapters are rather more technical, and a basic understanding of astronomy is necessary to enjoy them.

Modern Theories of the Universe (Dover, 1994), by Michael J. Crowe, contains a very readable elementary account of the role of spectroscopy in the development of the distance scale. The discussion includes much historical background and selections from original sources.

SEVEN: ISLANDS IN THE SKY

A complete list of the objects from the Messier catalog can be found in many places, such as in an appendix in M. W. Friedlander, *Astronomy from Stonehenge to Quasars* (Prentice-Hall, 1985).

Crowe's *Modern Theories of the Universe* (Dover, 1994) is the best source for a discussion of most of the subjects in this chapter. Crowe includes selections from original papers by Herschel, Shapley, Curtis, and Hubble, with annotations to guide the reader.

Much has been written on the Shapley–Curtis "Great Debate." Robert W. Smith's excellent book on the subject, *The Expanding Universe: Astronomy's "Great Debate"* (Cambridge University Press, 1982), has a long list of references. The account here is drawn from the texts of the papers published in a 1921 issue of the *Bulletin of the National Research Council*, vol. 2, p. 171 (1921). However, the written versions of the debate were prepared well after the actual debate, and what actually occurred at 8:15 PM on April 26, 1920, was quite different from the written versions. For instance, Shapley and Curtis aimed their talks at very different audiences. Shapley's presentation was for the general public, whereas Curtis's was technical and aimed at the professional astronomer. Many such details about what actually happened can be found in an article by Michael A. Hoskin, "The 'Great Debate': What Really Happened," *Journal for the History of Astronomy*, vol. 7, p. 169 (1982).

The simple question "Who won the debate?" is not so simple to answer. Curtis was correct about island universes, Shapley overestimated the size of the Milky Way while Curtis underestimated it, with Shapley somewhat closer to the true value. On reading the written versions of the debate, I concluded that if one awarded points as in a debating match, Shapley would have won. However, many eyewitnesses to the actual debate believed Curtis, an experienced public speaker, did a better job with the presentation than young Shapley, still only in his mid-thirties. Perhaps the most important thing Shapley had on his mind was to impress the few members of the audience from the Harvard Observatory visiting committee, who were looking him over for the vacant position of director (Shapley eventually was offered the position). Perhaps the basic lesson from all of this is that debates have never settled scientific issues, and never will.

- p. 172: "I do not understand how to quote..."

Leonardo's notebooks.

- p. 173: "...[The Milky Way] is in fact nothing but a congeries..."

Galileo in *The Starry Messenger.*

- p. 179: "Seeing is in some respects an art..."

This can be found in *The Scientific Papers of Sir William Herschel,* by J. L. E. Dreyer. It is quoted in Crowe.

- p. 181: "We may also have surmised nebulae..."

William Herschel, *Philosophical Transactions of the Royal Society* (1811). It may be found in Dreyer's collection of Herschel's papers, and it is reprinted in Crowe.

- p. 190–91: "Another consequence of the conclusion..."
  "There is a unity and internal agreement..."

From the written proceedings of the debate (reference given above).

EIGHT: THE EXPANDING FOG

Einstein's discovery of general relativity and its application to cosmology is discussed in the wonderful biography *Subtle Is the Lord—the Sci-*

ence and Life of Albert Einstein (Oxford University Press, 1982), by Abraham Pais.

Some of the original papers of Einstein, Lemaître, Friedmann, Hubble, and others who played important roles in the development of cosmology in the first part of the twentieth century are reprinted in *Cosmological Constants* (Columbia University Press, 1986), by Jeremy Bernstein and Gerald Feinberg. The accompanying annotations by the authors are a great help in placing the historical papers in a modern context.

The experiment of Buijs-Ballot (often spelled Buys-Ballot, sometimes just Ballot) has been described incorrectly in many places, usually exaggerating the number of musicians involved. The scientific description of the experiment is in the German journal *Annalen der Physik und Chemie*, vol. 66, p. 21 (1845), while the popular description in serialized form appeared in the Dutch music magazine *Caecilia* during the summer months of 1845.

Information about Hubble has been drawn from the article by N. U. Mayall in *Biographical Memoirs* of the National Academy of Sciences (Columbia University Press, 1970), vol. XLI, p. 175; "Young Edwin Hubble," by Donald E. Osterbrock, Ronald S. Brashear, and Joel A. Gwinn, in the magazine *Mercury* (January/February 1990); by Alexander S. Sharov and Igor D. Novikov, *Edwin Hubble: The Discoverer of the Big Bang Universe*, (Cambridge University Press, 1993). The information in many accounts of Hubble's life is drawn from Mayall's article, which is regarded as less than reliable.

Once again, Michael Crowe's book *Modern Theories of the Universe* (Dover, 1994) is highly recommended reading for the subjects in this chapter.

The most complete biography to date is *Edwin Hubble* (Farrar, Straus, Giroux, 1995) by Gale E. Christianson.

- p. 206: "I shall conduct the reader over the road..."
Einstein, "Cosmological Considerations on the General Theory of Relativity," *Proceedings of the Prussian Academy of Sciences*, vol. X, p. 142 (1917).[1]

- p. 208: "...the lines in the spectra of very distant stars..."
de Sitter, "On Einstein's Theory of Gravitation, and Its Astronomical

[1] This paper is reprinted in Bernstein and Feinberg.

Consequences," *Monthly Notices of the Royal Astronomical Society*, vol. 78, p. 3 (1917).[2]

• p. 212: "I have in an earlier note..."
Einstein, "Comments on the Work of A. Friedmann," *Zeitschrift für Physik*, vol. 16, p. 228 (1922).[3]

• p. 216: "Personally, [Hubble] is a man of the finest type..."
Quoted in the article by Osterbrock, Brashear, and Gwinn.

• p. 225: "The outstanding feature, however, is the..."
Hubble, "A Relation Between Distance and Radial Velocity Among Extra-galactic Nebulae," *Proceedings of the National Academy of Science*, vol. 15, p. 168 (1929).

• p. 233: "**Salviati:** I might very reasonably dispute..."
Galileo, in the *Dialog*, Day 3. I have quoted from the translation by Stillman Drake (p. 315).

NINE: A MATTER OF DEGREES

The story of the discovery of the microwave background radiation has been told many times. Two particularly good sources are Jeremy Bernstein, *Three Degrees Above Zero: Bell Labs in the Information Age* (Scribner's, 1984), and Robert M. Wilson's paper in *Modern Cosmology in Retrospective*, (Cambridge University Press, 1990), edited by B. Bertotti, R. Balbinot, S. Bergia, and A. Messina. Up-to-date information about the background radiation can be found in George Smoot and Keay Davidson, *Wrinkles in Time* (Morrow, 1993).

Details of the life of Gamow can be found in his amusing autobiography, *My World Line: An Informal Autobiography* (Viking Press, 1970). (Of course, experienced readers know better than to believe everything found in autobiographies.) The early work of Alpher, Herman, and Gamow was described in the article by Ralph A. Alpher and Robert Herman in *Modern Cosmology in Retrospective.*

Some noteworthy technical papers are the discovery paper of Penzias and Wilson, "A Measurement of Excess Antenna Temperature at 4080 Mc/s," *Astrophysical Journal*, vol. 142, p. 115 (1965), and "Measurement of the Flux Density of Cas A at 4080 Mc/s," *Astrophysical*

[2] This paper is reprinted in Bernstein and Feinberg.

[3] This paper is reprinted in Bernstein and Feinberg.

Journal, vol. 142, p. 1149 (1965); R. A. Alpher, J. W. Follin, Jr., and R. Herman, "Physical Conditions in the Initial Stages of the Expanding Universe," *Physical Review*, vol. 92, p. 1347 (1953).

TEN: PRIMORDIAL SOUP

There are several recent books about modern cosmology and the early universe. The best-known popular treatment is *The First Three Minutes: A Modern View of the Origin of the Universe*, (Basic Books, 1977), by Steven Weinberg.

Many of my friends and colleagues have written popular accounts of the big bang (I am sure to hear from those inadvertently omitted): Joseph Silk, *The Big Bang* (Freeman, 1989); John D. Barrow and Joseph Silk, *The Left Hand of Creation* (Basic Books, 1983); several books by Paul C. W. Davies, including *Superforce* (Simon and Schuster, 1984); Heinz R. Pagles, *The Cosmic Code* (Bantam Books, 1983); several books by James Trefil, including *The Moment of Creation* (Scribner's, 1983). A more technical treatment of particle physics and its applications to the early universe is a reprint volume of *Scientific American* articles called *Particle Physics in the Cosmos* (W. H. Freeman, 1989), edited by Richard A. Carrigan, Jr., and W. Peter Trower.

There are many popular discussions about elementary particle physics. A selection of popular books on the subject include Steven Weinberg, *The Discovery of Elementary Particles* (Scientific American Library, 1990); Frank Close, Michael Martin, and Christine Sutton, *The Particle Explosion* (Oxford University Press, 1987); Christine Sutton, *The Particle Connection* (Simon and Schuster, 1984); Frank Wilczek and Betsy Devine, *Longing for the Harmonies* (W. W. Norton, 1988); Anthony Zee, *Fearful Symmetry: The Search for Beauty in Modern Physics* (Collier Books, 1989).

A very recent book about high-energy physics is *The Particle Garden* (Addison-Wesley, 1995), by Gordon L. Kane. *Dreams of a Final Theory* (Pantheon, 1992), by Steven Weinberg is about the intellectual adventure of high-energy physics.

For a personal tale of the joy of discovery in high-energy physics, see *The God Particle: If the Universe Is the Answer, What Is the Question?* (Houghton Mifflin, 1993), by Leon M. Lederman.

Finally, a discussion of creation myths can be found in the first chapter of the book *The Dancing Universe* (Dutton, forthcoming), by Marcelo Gleiser.

- p. 267: "Some foolish men declare Creator made the world..."

Marcelo Gleiser, *The Dancing Universe.*

- p. 274: "The world was never finished until P'an Ku died..."

Marcelo Gleiser, *The Dancing Universe.*

- p. 276: "In advocating and fighting for the Copernican theory..."

Albert Einstein, in Stillman Drake's translation of Galileo's *Dialog.* See the sources quoted for Chapter 4.

ELEVEN: THE RAW EDGE

Two books about invisible matter are L. M. Krauss, *The Fifth Essence: The Search for Dark Matter in the Universe* (Basic Books, 1989), and Michael Riordan and David Schramm, *The Shadows of Creation: Dark Matter and the Structure of the Universe* (W. H. Freeman, 1990).

Permissions, Sources, and Credits

Page iii:	Reprinted with permission from *The Glorious Constellations* by Giuseppe Maria Sesti (Abrams, New York, 1991).
Page 4:	From *Uranometria*, Johannes Bayer, 1603. Reprinted with permission from *The Glorious Constellations* by Giuseppe Maria Sesti (Abrams, New York, 1991).
Page 5:	California Institute of Technology.
Page 6:	California Institute of Technology.
Page 11:	Camille Flammarion, 1888. Reprinted with permission from *The Glorious Constellations* by Giuseppe Maria Sesti (Abrams, New York, 1991).
Page 16:	From *Uranometria*, Johannes Bayer, 1603. Reprinted with permission from *The Glorious Constellations* by Giuseppe Maria Sesti (Abrams, New York, 1991).
Page 20:	From *Epistolarum Astronomicarum*, Tycho Brahe, 1596.

Page 22: From *Astronomiæ Instauratæ Mechanica*, Tycho Brahe, 1598.

Page 24: From *Astronomiæ Instauratæ Mechanica*, Tycho Brahe, 1598.

Page 26: From *Astronomiæ Instauratæ Mechanica*, Tycho Brahe, 1598.

Page 28: Original illustration by Rocky Kolb.

Page 29: Original illustration by Rocky Kolb.

Page 32: Original illustration by Rocky Kolb.

Page 34: Original illustration by Rocky Kolb.

Page 38: From *Astronomiæ Instauratæ Mechanica*, Tycho Brahe, 1598.

Page 48: Mary Lea Shane Archives of the Lick Observatory.

Page 52: Original illustration by Rocky Kolb.

Page 53: Original illustration by Rocky Kolb.

Page 53: From *Mysterium Cosmographicum*, Johannes Kepler, 1596. Reproduced in *Gesammelte Werke.* Im Auftrag der Deutschen Forschungsgemeinschaft und der Wissenschaften begründet von Walther von Dyck und Max Casper, herausgegeben von Franz Hammer. C. H. Beck'sche Verlagsbuchhandlung, München.

Page 58: Original illustration by Rocky Kolb.

Page 60: Original illustration by Rocky Kolb.

Page 61: Original illustration by Rocky Kolb.

Page 62: Original illustration by Rocky Kolb.

Page 67: From *Harmonice Mundi*, Johannes Kepler, 1619. Reproduced in *Gesammelte Werke.* Im Auftrag der Deutschen Forschungsgemeinschaft und der Wissenschaften begründet von Walther von Dyck und Max Casper, herausgegeben von Franz Hammer. C. H. Beck'sche Verlagsbuchhandlung, München.

Page 68: Original illustration by Rocky Kolb.

Page 81: Yerkes Observatory.

Page 85: Yerkes Observatory.

Page 88: From *Robert Bellarmine, Saint and Scholar*, James Broderick, S.J. (Newman Press, 1961).

Page 95: Original illustration by Rocky Kolb.

Page 98: Portrait by Pietro da Cortona, Museo di Roma and Galleria Capitolina, Rome.

Page 100: From *Dialogo Sopra I Due Massimi Sistemi del Mondo*, Galileo, 1632.

Page 117: Yerkes Observatory.

Page 120: (Newton) By permission of the Syndics of Cambridge University Library. Photograph courtesy of R. S. Westfall; (Leonardo) The Royal Collection © Her Majesty Queen Elizabeth II.

Page 123: Property of Biblieoteca Ambrosiana. All rights reserved. No reproductions allowed.

Page 127: (Descartes) Frans Hals, Statens Museum for Kunst, Copenhagen. Reproduced by permission; (Gilbert) From *On the Magnet*, William Gilbert, 1600.

Page 137: Original art by Angela Gonzales.

Page 143: Original illustration by Rocky Kolb.

Page 144: Original illustrations by Rocky Kolb.

Page 147: Original illustration by Rocky Kolb.

Page 149: Original illustration by Rocky Kolb.

Page 150: Original illustration by Rocky Kolb.

Page 156: Original illustration by Rocky Kolb.

Page 160: From *Uranometria*, Johannes Bayer, 1603. Reprinted with permission from *The Glorious Constellations* by Giuseppe Maria Sesti (Abrams, New York, 1991).

Page 161: Yerkes Observatory.

Page 165: Original illustration by Rocky Kolb.

Page 166: H. N. Russell, *Popular Astronomy* **22**, 275 (1914).

Page 167: Reprinted with permission from W. Gleise, "Hertsprung-Russell Diagrams and Color-Luminosity Diagrams for the Stars Nearer than Twenty-Two Parsecs," in *The HR Diagram* (IAU Symp. 80), A.. G. Davis Phillip and D. S. Hayes, eds. (Reidel, 1978), pp.79–88.

Page 171:	The Royal Collection © Her Majesty Queen Elizabeth II.
Page 175:	(M42) Yerkes Observatory; (M57) Yerkes Observatory; (M3) Yerkes Observatory; (M66) U. S. Naval Observatory.
Page 178:	Yerkes Observatory.
Page 182:	Yerkes Observatory.
Page 183:	(M51—Rosse's illustration) From *Philosophical Transactions of the Royal Society of London*, 1850; (M51—modern picture) Yerkes Observatory.
Page 185:	Courtesy of Steve Kent, Fermilab.
Page 188:	(Shapley) Yerkes Observatory; (Curtis) Mary Lea Shane Archives of the Lick Observatory.
Page 195:	The Observatories of the Carnegie Institution of Washington.
Page 199:	Original art by Angela Gonzales.
Page 205:	Courtesy of Stephane Colombi.
Page 214:	From *Cap and Gown*—University of Chicago yearbook, 1910.
Page 215:	From *The Senior Blotter*—New Albany, Indiana High School yearbook, 1914. Photograph courtesy of Donald E. Osterbrock.
Page 218:	The Observatories of the Carnegie Institution of Washington.
Page 226:	Reproduced by permission of the Huntington Library.
Page 228:	Original illustration by Rocky Kolb.
Page 229:	Original illustration by Rocky Kolb.
Page 239:	AT & T Archives.
Page 240:	From *Modern Cosmology in Retrospect*, edited by B. Bertotti, R. Balbonit, S. Bergia, and A. Messina, (Cambridge Univ. Press, 1990). Reprinted with the permission of Cambridge University Press. Photograph courtesy of Robert W. Wilson.
Page 245:	(Gamow) AIP Emilio Segrè Visual Archives, *Physics Today* Collection; (Lemaîre) Yerkes Observatory.
Page 251:	Courtesy of Ralph Alpher.

Page 254: Original illustration by Rocky Kolb.

Page 255: Courtesy NRAO/AUI.

Page 257: From *Modern Cosmology in Retrospect*, edited by B. Bertotti, R. Balbonit, S. Bergia, and A. Messina, (Cambridge Univ. Press, 1990). Reprinted with the permission of Cambridge University Press. Photograph courtesy of Robert W. Wilson.

Page 263: From *Astronomiæ Instauratæ Mechanica*, Tycho Brahe 1596.

Page 264: Courtesy of The Adler Planetarium and Astronomy Museum, Chicago, Illinois.

Page 265: Courtesy of Fermilab Visual Media Services.

Page 277: Courtesy of Fermilab Visual Media Services.

Page 278: Courtesy of Thom Edel and Jim Shultz.

Page 286: From the *Encyclopedia Americana*, 1995 Edition. Copyright 1995 by Grolier Incorporated. Reprinted by permission.

Page 296: Original illustration by Rocky Kolb.

Page 298: Original illustration by Rocky Kolb.

Page 301: Original illustration by Rocky Kolb.

Page 303: Original illustration by Rocky Kolb.

Page 305: Original illustration by Rocky Kolb.

Acknowledgments

This book would not have been written without the help and encouragement of many people. Comments about the first fledgling chapters by Kate Metropolis, Judy Jackson, Vicki Jennings, and Jack Repcheck encouraged me to write more. Hans Kautsky kindly translated a paper of Christopher H. D. Buijs-Ballot from a nineteenth-century German physics journal. Alexandra Battaglia Mayer supplied information about the Roman church of Santa Maria Sopra Minerva. Reverend George V. Coyne, S. J., took time from his duties as director of the Vatican Observatory to recommend references about Saint Bellarmine's life and interactions with Galileo. Richard Dreiser of Yerkes Observatory, and Reidar Hahn and Sheila Colson of Fermilab provided invaluable assistance with the illustrations. My University of Chicago colleague, the historian Noel Swerdlow, never seemed to tire of my questions about Tycho, Kepler, and Galileo. Many scientific collaborators were remarkably patient while I was distracted from other scientific matters during the writing of this book. Angela Gonzales is always generous with her work and views as an artist.

Marcelo Gleiser read every chapter and offered many suggestions that resulted in a better book. Obrigadão, cara!

Special thanks to Leon Lederman for graciously taking time from a busy life to write a foreword.

The transformation of a manuscript into a book was possible because of the care and attention of my editor, Jeff Robbins, the production supervisor, Lynne Reed, and my literary agents, Katinka Matson and John Brockman.

The deepest gratitude is reserved for my family for allowing a year of nights and weekends to pass while I typed away in my study.

This book was written while listening to the operas of Giuseppe Verdi. I would like to acknowledge the spirit that drove him to write his operas.

Index